PyTorch Recipes

A Problem-Solution Approach to Build, Train and Deploy Neural Network Models

Second Edition

Pradeepta Mishra

Apress®

PyTorch Recipes: A Problem-Solution Approach to Build, Train and Deploy Neural Network Models

Pradeepta Mishra
Bangalore, Karnataka, India

ISBN-13 (pbk): 978-1-4842-8924-2   ISBN-13 (electronic): 978-1-4842-8925-9
https://doi.org/10.1007/978-1-4842-8925-9

Managing Director, Apress Media LLC: Welmoed Spahr
Acquisitions Editor: Celestin Suresh John
Development Editor: James Markham
Coordinating Editor: Mark Powers
Copy Editor: Mary Behr

Cover designed by eStudioCalamar

Cover image by Marek Piwinicki on Unsplash (www.unsplash.com)

Distributed to the book trade worldwide by Apress Media, LLC, 1 New York Plaza, New York, NY 10004, U.S.A. Phone 1-800-SPRINGER, fax (201) 348-4505, e-mail orders-ny@springer-sbm.com, or visit www.springeronline.com. Apress Media, LLC is a California LLC and the sole member (owner) is Springer Science + Business Media Finance Inc (SSBM Finance Inc). SSBM Finance Inc is a **Delaware** corporation.

For information on translations, please e-mail booktranslations@springernature.com; for reprint, paperback, or audio rights, please e-mail bookpermissions@springernature.com.

Apress titles may be purchased in bulk for academic, corporate, or promotional use. eBook versions and licenses are also available for most titles. For more information, reference our Print and eBook Bulk Sales web page at www.apress.com/bulk-sales.

Any source code or other supplementary material referenced by the author in this book is available to readers on GitHub (https://github.com/Apress). For more detailed information, please visit www.apress.com/source-code.

Printed on acid-free paper

*I would like to dedicate this book to my dear parents,
my lovely wife, Prajna, and my daughters, Priyanshi (Aarya)
and Adyanshi (Aadya). This work would not have been possible
without their inspiration, support, and encouragement.*

Table of Contents

About the Author

Pradeepta Mishra is an AI leader, an experienced data scientist, and an artificial intelligence architect. He currently heads NLP, ML, and AI initiatives for five products at FOSFOR by LTI, a leading-edge innovator in AI and machine learning based out of Bangalore, India. He has expertise in designing artificial intelligence systems for performing tasks such as understanding natural language and recommendations based on natural language processing. He has filed 12 patents as an inventor and has authored and co-authored five books: *R Data Mining Blueprints* (Packt Publishing, 2016), *R: Mining Spatial, Text, Web, and Social Media Data* (Packt Publishing, 2017), *PyTorch Recipes First Edition* (Apress, 2019), and *Practical Explainable AI Using Python* (Apress, 2022). There are two courses available on Udemy based on these books.

Pradeepta presented a keynote talk on the application of bidirectional LSTM for time series forecasting at the 2018 Global Data Science Conference. He delivered a TEDx Talk titled "Can Machines Think?" on the power of artificial intelligence in transforming industries and changing job roles across industries. He has also delivered more than 150 tech talks on data science, machine learning, and artificial intelligence at various meetups, technical institutions, universities, and community forums. He is on LinkedIn at www.linkedin.com/in/pradeepta/ and Twitter at @pradmishra1.

About the Technical Reviewer

 **Chris Thomas** is a UK-based consultant specializing in artificial intelligence and machine learning research and development. As a professional member of the Institute of Analysts and Programmers, Chris's knowledge is based on a career as a technical professional with over 20 years of experience in the public, semiconductor, finance, utilities, and marketing sectors.

Acknowledgments

I would like to thank my wife, Prajna, for her continuous inspiration and support, and for sacrificing her weekends just to sit alongside me to help me complete this book; and my daughters, Aarya and Aadya, for being patient all through my writing time.

A big thank you to Celestin Suresh John and Mark Powers for fast-tracking the whole process and helping me and guiding me in the right direction.

Introduction

The development of artificial intelligent products and solutions has recently become a norm, so the demand for graph theory–based computational frameworks is on the rise. Making the deep learning models work in real-life applications is possible when the modeling framework is dynamic, flexible, and adaptable to other frameworks.

PyTorch is a recent entrant to the league of graph computation tools/programming languages. Addressing the limitations of previous frameworks, PyTorch promises a better user experience in the deployment of deep learning models and the creation of advanced models using a combination of convolutional neural networks, recurrent neural networks, LSTMs, and deep neural networks.

PyTorch was created by Facebook's Artificial Intelligence Research division, which seeks to make the model development process simple, straightforward, and dynamic so that developers do not have to worry about declaring objects before compiling and executing the model. It is based on the Torch framework and is an extension of Python.

This book is intended for data scientists, natural language processing engineers, artificial intelligence solution developers, existing practitioners working on graph computation frameworks, and researchers of graph theory. This book will get you started with understanding tensor basics and computation. You'll learn how to perform arithmetic-based operations, matrix algebra, and statistical distribution-based operations using the PyTorch framework.

Chapters 3 and 4 provide detailed descriptions of neural network basics. Advanced neural networks such as convolutional neural networks, recurrent neural networks, and LSTMs are explored. You will be able to implement these models using PyTorch functions.

Chapters 5 and 6 discuss fine-tuning the models, hyper parameter tuning, and the refinement of existing PyTorch models in production. You will learn how to choose the hyper parameters to fine-tune the model.

In Chapter 7, natural language processing is explained. The deep learning models and their applications in natural language processing and artificial intelligence is one of the most demanding skill sets in the industry. You will be able to benchmark the execution and performance of a PyTorch implementation in deep learning models to execute and process natural language. You will compare PyTorch with other graph computation–based deep learning programming tools.

Source Code

Go to github.com/apress/pytorch-recipes-2e for all source code and other supplementary material referenced by the author.

CHAPTER 1

Introduction to PyTorch, Tensors, and Tensor Operations

PyTorch has continued to evolve as a larger framework for writing dynamic models. Because of this, it is very popular among data scientists and data engineers for deploying large-scale deep learning frameworks. This book provides a structure for experts in terms of handling activities while working on practical data science problems. As evident from applications that we use in our day-to-day lives, there are layers of intelligence embedded within a product's features. These features are enabled to provide a better experience and better services to users.

The world is moving toward artificial intelligence. The real potential of artificial intelligence is achieved through developing trainable systems. Machine learning is suitable for low-dimensional data and for small volumes of data. Deep learning is suitable when the data dimension is huge and the training data is also high in volume.

PyTorch is the most optimized high-performance tensor library for computation of deep learning tasks on GPUs (graphics processing units) and CPUs (central processing units). The main purpose of PyTorch is to enhance the performance of algorithms in large-scale computing environments. PyTorch is a library based on Python and the Torch tool provided by Facebook's Artificial Intelligence Research group, which performs scientific computing.

NumPy-based operations on a GPU are not efficient enough to process heavy computations. Static deep learning libraries are a bottleneck for bringing flexibility to computations and speed. From a practitioner's point of view, PyTorch tensors are very similar to the N-dimensional arrays of a NumPy library based on Python. The PyTorch library provides bridge options for moving a NumPy array to a tensor array, and vice versa, in order to make the library flexible across different computing environments.

© Pradeepta Mishra 2023
P. Mishra, *PyTorch Recipes*, https://doi.org/10.1007/978-1-4842-8925-9_1

The use cases where it is most frequently used include tabular data analysis, natural language processing, image processing, computer vision, social media data analysis, and sensor data processing. Although PyTorch provides a large collection of libraries and modules for computation, three modules are very prominent.

- `Autograd`: This module provides functionality for automatic differentiation of tensors. A recorder class in the program remembers the operations and retrieves those operations with a trigger called `backward` to compute the gradients. This is immensely helpful in the implementation of neural network models.

- `Optim`: This module provides optimization techniques that can be used to minimize the error function for a specific model. Currently, PyTorch supports various advanced optimization methods, which includes Adam, stochastic gradient descent (SGD), and more.

- NN: NN stands for *neural network* model. Manually defining the functions, layers, and further computations using complete tensor operations is very difficult to remember and execute. We need functions that automate the layers, activation functions, loss functions, and optimization functions, and provides a layer defined by the user so that manual intervention can be reduced. The NN module has a set of built-in functions that automates the manual process of running a tensor operation.

Industries in which artificial intelligence is applied include banking, financial services, insurance, health care, manufacturing, retail, clinical trials, and drug testing. Artificial intelligence involves classifying objects, recognizing the objects to detect fraud, and so forth. Every learning system requires three things: input data, processing, and an output layer. Figure 1-1 explains the relationship between these three topics. If the performance of any learning system improves over time by learning from new examples or data, it is called a *machine learning system*. When a machine learning system becomes too difficult to reflect reality, it often requires a deep learning system.

In a deep learning system, more than one hidden layer of a learning algorithm is deployed. In machine learning, we think of supervised, unsupervised, semisupervised, and reinforcement learning systems. A supervised machine-learning algorithm is one where the data is labeled with classes or tagged with outcomes. We show the machine the input data with corresponding tags or labels. The machine identifies the relationship with a function. Please note that this function connects the input to the labels or tags.

In unsupervised learning, we show the machine only the input data and ask the machine to group the inputs based on association, similarities or dissimilarities, and so forth.

In semisupervised learning, we show the machine input features and labeled data or tags and we ask the machine to predict the untagged outcomes or labels.

In reinforcement learning, we introduce a reward and penalty mechanism, where each and every policy goes through a round of iteration and usually a correct action is rewarded and an incorrect action is penalized to maintain the status of the policy.

In all of these examples of machine learning algorithms, we assume that the dataset is small, because getting massive amounts of tagged data is a challenge, and it takes a lot of time for machine learning algorithms to process large-scale matrix computations. Since machine learning algorithms are not scalable for massive datasets, we need deep learning algorithms.

Figure 1-1 shows the relationships among artificial intelligence, machine learning, and deep learning. Natural language is an important part of artificial intelligence. We need to develop systems that understand natural language and provide responses to the agent. Let's take an example of machine translation where a sentence in language 1 (French) can be converted to language 2 (English) and vice versa. To develop such a system, we need a large collection of English-French bilingual sentences. The corpus requirement is very large, as all language nuances need to be covered by the model.

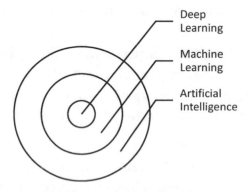

Figure 1-1. *Relationships among ML, DL, and AI*

After preprocessing and feature creation, you can observe hundreds of thousands of features that need to be computed to produce output. To train a machine learning supervised model would take months to run and to produce output. To achieve scalability in this task, we need deep learning algorithms, such as a recurrent neural network. This is how the artificial intelligence is connected to deep learning and machine learning.

There are various challenges in deploying deep learning models that require large volumes of labeled data, faster computing machines, and intelligent algorithms. The success of any deep learning system requires well-labeled data and better computing machines because the smart algorithms are already available.

The following are various use cases where a deep learning implementation is very effective:

- Speech recognition

- Video analysis

- Anomaly detection from videos

- Natural language processing

- Machine translation

- Speech-to-text conversion

The development of the NVIDIA GPU for processing large-scale data is another path-breaking innovation. The programming language that is required to run in a GPU environment requires a different programming framework. Two major frameworks are very popular for implementing graphical computing: TensorFlow and PyTorch. In this book, I discuss PyTorch as a framework to implement data science algorithms and make inferences.

The major frameworks for graph computations include PyTorch, TensorFlow, and MXNet. PyTorch and TensorFlow compete with each other in neurocomputations. TensorFlow and PyTorch are similar in terms of performance; however, the real differences are known only when we benchmark a particular task. Concept-wise there are certain differences.

- In TensorFlow, we must define the tensors, initialize the session, and keep placeholders for the tensor objects; however, we do not have to do these operations in PyTorch.

- In TensorFlow, let's consider sentiment analysis as an example. Input sentences are tagged with positive or negative tags. If the input sentence's length is not equal, then we set the maximum sentence

length and add zero to make the length of other sentences equal,
so that the recurrent neural network can function; however, this is
a built-in functionality in PyTorch, so we do not have to define the
length of the sentences.

- In PyTorch, the debugging is easy and simple, but it is a difficult task
  in TensorFlow.

- In terms of data visualization, model deployment is definitely better
  in TensorFlow; however, PyTorch is evolving, and we expect to
  eventually see the same functionality in it in the future.

TensorFlow has definitely undergone many changes to reach a stable state. PyTorch
has really come a long way and provides a stable deep learning framework. PyTorch
becomes a standard for all large scale transformer based models, available on a hugging
face platform.

What Is PyTorch?

PyTorch is a machine learning and deep learning tool developed by Facebook's
Artificial Intelligence Research division to process large-scale image analysis, including
object detection, segmentation, and classification. It is not limited to these tasks,
however. It can be used with other frameworks to implement complex algorithms. It
is written using Python and the C++ language. To process large-scale computations
in a GPU environment, the programming languages should be modified accordingly.
PyTorch provides a great framework to write functions that automatically run in a GPU
environment.

PyTorch Installation

Installing PyTorch is quite simple. In Windows, Linux, or macOS, it is very simple to
install if you are familiar with the Anaconda and Conda environments for managing
packages. The following steps describe how to install PyTorch in Windows/macOS/
Linux environments.

1. Open the Anaconda navigator and go to the environment page, as
   displayed in Figure 1-2.

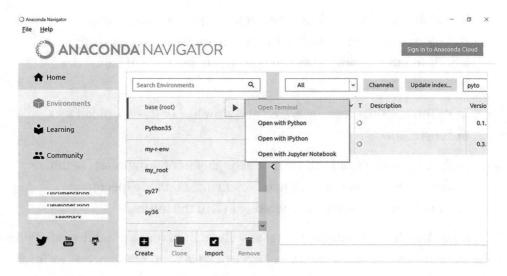

Figure 1-2. *Anaconda Navigator for Installing PyTorch*

2.   Open the terminal and type the following:

```
conda install -c peterjc123 pytorch
```

3.   Launch Jupyter and open the IPython Notebook.

4.   Type the following command to check whether the PyTorch is installed or not:

```
from __future__ import print_function
import torch
```

5.   Check the version of PyTorch.

```
torch.version.__version__
1.12.1+cu113
```

This installation process was done using a Microsoft Windows machine. The process may vary by operating system, so please use the following URLs for any issue regarding installation and errors.

There are two ways to install it: the Conda (Anaconda) library management or the Pip3 package management framework. Also, installations for a local system (such as macOS, Windows, or Linux) and a cloud machine (such as Microsoft Azure, AWS, and GCP) are different. To set up according to your platform, please follow the official PyTorch installation documents at `https://PyTorch.org/get-started/cloud-partners/`.

PyTorch has various components.

- Torch has functionalities similar to NumPy with GPU support.

- Autograd's `torch.autograd` provides classes, methods, and functions for implementing automatic differentiation of arbitrary scalar valued functions. It requires minimal changes to the existing code. You only need to declare `class:'Tensor's`, for which gradients should be computed with the `requires_grad=True` keyword.

- NN is a neural network library in PyTorch.

- Optim provides optimization algorithms that are used for the minimization and maximization of functions.

- Multiprocessing is a useful library for memory sharing between multiple tensors.

- Utils has utility functions to load data; it also has other functions.

Now you are ready to proceed with the chapter.

Recipe 1-1. Using Tensors
Problem

The data structure used in PyTorch is graph based and tensor based, so it is important to understand basic operations and defining tensors, such as indexing, reshaping, and changing data types.

Solution

The solution to this problem is practicing on the tensors and its operations, which includes many examples that use various operations. Although it is assumed that you are familiar with PyTorch and Python basics, a refresher on PyTorch is essential to create interest among new users.

How It Works

Let's have a look at the following examples of tensors and tensor operation basics, including mathematical operations.

The x object is a list. You can check whether an object in Python is a tensor object by using the following syntax. Typically, the is_tensor function checks and is_storage function checks whether the object is stored as tensor object.

```
x = [12,23,34,45,56,67,78]
```

```
# Scalar
scalar = torch.tensor(10)
scalar
tensor(10)
```

```
scalar.ndim
0
```

```
scalar.item()
10
```

```
#vector
vector = torch.tensor([5,5])
vector
```

```
tensor([5, 5])
```

```
vector.ndim
1
```

```
vector.shape
torch.Size([2])
```

```
# Matrix
matrix = torch.tensor([[4, 5],
                       [10, 110]])
matrix
```

```
tensor([[ 4, 5], [ 10, 110]])
```

```
matrix.ndim
2

matrix.shape
torch.Size([2, 2])

# Tensor is multidimensional
tensor = torch.tensor([[[4,64, 5,4],
                        [10,20,30, 110],
                        [45,34,67,40],
                        [56,67,89,90]]])
tensor

tensor([[[ 4, 64, 5, 4], [ 10, 20, 30, 110], [ 45, 34, 67, 40], [ 56, 67,
89, 90]]])

tensor.ndim
3

tensor.shape
torch.Size([1, 4, 4])

tensor.dtype
torch.int64

tensor.device
device(type='cpu')

torch.is_tensor(x)
False

torch.is_storage(x)
False
```

Now, let's create an object that contains random numbers from Torch, similar to NumPy library. You can check the tensor and storage type.

```
y = torch.randn(1,2,3,4,5)

torch.is_tensor(y)
True
```

9

```
torch.is_storage(y)
False
```

```
torch.numel(y) # the total number of elements in the input Tensor
120
```

The y object is a tensor; however, it is not stored. To check the total number of elements in the input tensor object, the numerical element function can be used. The following script is another example of creating zero values in a 2D tensor and counting the numerical elements in it:

```
torch.zeros(4,4)
```

```
tensor([[0., 0., 0., 0.], [0., 0., 0., 0.], [0., 0., 0., 0.], [0., 0.,
0., 0.]])
```

```
torch.numel(torch.zeros(4,4))
```

```
torch.eye(3)
tensor([[1., 0., 0.], [0., 1., 0.], [0., 0., 1.]])
```

```
torch.eye(5)
tensor([[1., 0., 0., 0., 0.], [0., 1., 0., 0., 0.], [0., 0., 1., 0., 0.],
[0., 0., 0., 1., 0.], [0., 0., 0., 0., 1.]])
```

Like NumPy operations, the eye function creates a diagonal matrix, of which the diagonal elements have ones and off diagonal elements have zeros. The eye function can be manipulated by providing the shape option. The following example shows how to provide the shape parameter:

```
torch.eye(3,4)
tensor([[1., 0., 0., 0.], [0., 1., 0., 0.], [0., 0., 1., 0.]])
```

```
torch.eye(5,4)
tensor([[1., 0., 0., 0.], [0., 1., 0., 0.], [0., 0., 1., 0.], [0., 0., 0.,
1.], [0., 0., 0., 0.]])
```

```
type(x)
list
```

Linear space and points between the linear space can be created using tensor operations. Let's use an example of creating 25 points in a linear space starting from value 2 and ending with 10. Torch can read from a NumPy array format.

```
import numpy as np
x1 = np.array(x)
x1
array([12, 23, 34, 45, 56, 67, 78])
```

```
torch.from_numpy(x1)
tensor([12, 23, 34, 45, 56, 67, 78])
```

```
torch.linspace(2, 10, steps=25) #linear spacing
tensor([ 2.0000, 2.3333, 2.6667, 3.0000, 3.3333, 3.6667, 4.0000, 4.3333,
4.6667, 5.0000, 5.3333, 5.6667, 6.0000, 6.3333, 6.6667, 7.0000, 7.3333,
7.6667, 8.0000, 8.3333, 8.6667, 9.0000, 9.3333, 9.6667, 10.0000])
```

```
torch.linspace(-10, 10, steps=15)
tensor([-1.0000e+01, -8.5714e+00, -7.1429e+00, -5.7143e+00, -4.2857e+00,
-2.8571e+00, -1.4286e+00, -2.3842e-07, 1.4286e+00, 2.8571e+00, 4.2857e+00,
5.7143e+00, 7.1429e+00, 8.5714e+00, 1.0000e+01])
```

Like linear spacing, logarithmic spacing can be created.

```
torch.logspace(start=-10, end=10, steps=15) #logarithmic spacing
tensor([1.0000e-10, 2.6827e-09, 7.1969e-08, 1.9307e-06, 5.1795e-05,
1.3895e-03, 3.7276e-02, 1.0000e+00, 2.6827e+01, 7.1969e+02, 1.9307e+04,
5.1795e+05, 1.3895e+07, 3.7276e+08, 1.0000e+10])
```

```
torch.ones(4)
tensor([1., 1., 1., 1.])
```

```
torch.ones(4,5)
tensor([[1., 1., 1., 1., 1.], [1., 1., 1., 1., 1.], [1., 1., 1., 1., 1.],
[1., 1., 1., 1., 1.]])
```

Random number generation is a common process in data science to generate or gather sample data points in a space to simulate structure in the data. Random numbers can be generated from a statistical distribution, any two values, or a predefined distribution. Like NumPy functions, the random number can be generated using the following example. Uniform distribution is defined as a distribution where each outcome has equal probability of happening; hence, the event probabilities are constant.

```
# random numbers from a uniform distribution between the values
# 0 and 1
torch.rand(10)
tensor([0.1408, 0.4445, 0.4251, 0.2663, 0.3743, 0.4784, 0.3760, 0.1876,
0.2151, 0.6876])
```

The following script shows how random numbers from two values, 0 and 1, are selected. The result tensor can be reshaped to create a (4,5) matrix. The random numbers from a normal distribution with arithmetic mean 0 and standard deviation 1 can also be created, as follows.

To select random values from a range of values using random permutation requires defining the range first. This range can be created by using the arrange function. When using the arrange function, you must define the step size, which places all the values in an equal distance space. By default, the step size is 1.

```
torch.rand(4, 5)
# random values between 0 and 1 and fillied with a matrix of
# size rows 4 and columns 5
tensor([[0.2733, 0.0302, 0.8835, 0.9537, 0.9662], [0.6296, 0.3106, 0.4029,
0.8133, 0.1697], [0.8578, 0.6517, 0.0440, 0.6197, 0.9889], [0.8614, 0.6288,
0.2158, 0.4593, 0.2444]])
```

```
#random numbers from a normal distribution,
#with mean =0 and standard deviation =1
torch.randn(10)
```

```
tensor([ 1.0115, -0.7502, 1.1994, 0.8736, 0.5633, -0.7702, 0.1826, -1.9931,
0.5159, 0.1521])
```

```
torch.randn(4, 5)
tensor([[ 0.3744, 2.1839, -1.8229, 1.0682, 1.5394], [ 0.9689, -1.3085,
-0.3300, 0.3960, -0.6079], [ 2.3485, 1.2880, 0.6754, -2.0426, -0.3121],
[-0.4897, -1.5335, 0.0467, -0.6213, 1.7185]])
```

```
#selecting values from a range, this is called random permutation
torch.randperm(10)

tensor([1, 6, 3, 2, 0, 8, 4, 5, 7, 9])

#usage of range function
torch.arange(10, 40,2) #step size 2

tensor([10, 12, 14, 16, 18, 20, 22, 24, 26, 28, 30, 32, 34, 36, 38])

torch.arange(10,40) #step size 1
tensor([10, 11, 12, 13, 14, 15, 16, 17, 18, 19, 20, 21, 22, 23, 24, 25, 26,
27, 28, 29, 30, 31, 32, 33, 34, 35, 36, 37, 38, 39])
```

To find the minimum and maximum values in a 1D tensor, argmin and argmax can be used. The dimension needs to be mentioned if the input is a matrix in order to search minimum values along rows or columns.

```
d = torch.randn(4, 5)
d
tensor([[ 1.0085, -0.8545, -0.6958, 1.6716, -0.0118], [ 0.2134, 1.1154,
-0.6426, -1.3651, -1.5724], [ 0.2452, 0.8356, 2.0297, -0.2397, 0.8560],
[ 0.9786, -0.8538, -0.6449, 0.3903, 1.5966]])

torch.argmin(d,dim=1)
tensor([1, 4, 3, 1])

torch.argmax(d,dim=1)
tensor([3, 1, 2, 4])
```

If it is either a row or column, it is a single dimension and is called a *1D tensor* or vector. If the input is a matrix, in which rows and columns are present, it is called a *2D tensor*. If it is more than two-dimensional, it is called a *multidimensional tensor*.

```
# create a 2dtensor filled with values as 0
torch.zeros(4,5)
tensor([[0., 0., 0., 0., 0.], [0., 0., 0., 0., 0.], [0., 0., 0., 0., 0.],
[0., 0., 0., 0., 0.]])
```

```
# create a 1d tensor filled with values as 0
torch.zeros(10)
tensor([0., 0., 0., 0., 0., 0., 0., 0., 0., 0.])
```

Now, let's create a sample 2D tensor and perform indexing and concatenation by using the concat operation on the tensors.

```
#indexing and performing operation on the tensors
x = torch.randn(4,5)
x
tensor([[-1.5343, -1.3533, -0.8621, -1.1674, -0.1114], [ 0.2790, 0.0463,
1.5364, -0.1287, 0.6379], [-0.4542, 0.5196, 0.2335, -0.5135, -0.6602],
[-0.6930, 0.0541, -0.8463, -0.4498, -0.0280]])
```

```
#concatenate two tensors
torch.cat((x,x))
tensor([[-1.5343, -1.3533, -0.8621, -1.1674, -0.1114], [ 0.2790, 0.0463,
1.5364, -0.1287, 0.6379], [-0.4542, 0.5196, 0.2335, -0.5135, -0.6602],
[-0.6930, 0.0541, -0.8463, -0.4498, -0.0280], [-1.5343, -1.3533, -0.8621,
-1.1674, -0.1114], [ 0.2790, 0.0463, 1.5364, -0.1287, 0.6379], [-0.4542,
0.5196, 0.2335, -0.5135, -0.6602], [-0.6930, 0.0541, -0.8463, -0.4498,
-0.0280]])
```

The sample x tensor can be used in 3D as well. Again, there are two different options to create three-dimensional tensors; the third dimension can be extended over rows or columns.

```
#concatenate n times based on array size
torch.cat((x,x,x))
tensor([[-1.5343, -1.3533, -0.8621, -1.1674, -0.1114], [ 0.2790, 0.0463,
1.5364, -0.1287, 0.6379], [-0.4542, 0.5196, 0.2335, -0.5135, -0.6602],
[-0.6930, 0.0541, -0.8463, -0.4498, -0.0280], [-1.5343, -1.3533, -0.8621,
-1.1674, -0.1114], [ 0.2790, 0.0463, 1.5364, -0.1287, 0.6379], [-0.4542,
0.5196, 0.2335, -0.5135, -0.6602], [-0.6930, 0.0541, -0.8463, -0.4498,
-0.0280], [-1.5343, -1.3533, -0.8621, -1.1674, -0.1114], [ 0.2790, 0.0463,
1.5364, -0.1287, 0.6379], [-0.4542, 0.5196, 0.2335, -0.5135, -0.6602],
[-0.6930, 0.0541, -0.8463, -0.4498, -0.0280]])
```

```
#concatenate n times based on array size, over column
torch.cat((x,x,x),1)
tensor([[-1.5343, -1.3533, -0.8621, -1.1674, -0.1114, -1.5343, -1.3533,
-0.8621, -1.1674, -0.1114, -1.5343, -1.3533, -0.8621, -1.1674, -0.1114], [
0.2790, 0.0463, 1.5364, -0.1287, 0.6379, 0.2790, 0.0463, 1.5364, -0.1287,
0.6379, 0.2790, 0.0463, 1.5364, -0.1287, 0.6379], [-0.4542, 0.5196, 0.2335,
-0.5135, -0.6602, -0.4542, 0.5196, 0.2335, -0.5135, -0.6602, -0.4542,
0.5196, 0.2335, -0.5135, -0.6602], [-0.6930, 0.0541, -0.8463, -0.4498,
-0.0280, -0.6930, 0.0541, -0.8463, -0.4498, -0.0280, -0.6930, 0.0541,
-0.8463, -0.4498, -0.0280]])

#concatenate n times based on array size, over rows
torch.cat((x,x),0)
tensor([[-1.5343, -1.3533, -0.8621, -1.1674, -0.1114], [ 0.2790, 0.0463,
1.5364, -0.1287, 0.6379], [-0.4542, 0.5196, 0.2335, -0.5135, -0.6602],
[-0.6930, 0.0541, -0.8463, -0.4498,  0.0280], [-1.5343, -1.3533, -0.8621,
-1.1674, -0.1114], [ 0.2790, 0.0463, 1.5364, -0.1287, 0.6379], [-0.4542,
0.5196, 0.2335, -0.5135, -0.6602], [-0.6930, 0.0541, -0.8463, -0.4498,
-0.0280]])

#how to split a tensor among small chunks
torch.arange(11).chunk(6)
        (tensor([0, 1]),
         tensor([2, 3]),
         tensor([4, 5]),
         tensor([6, 7]),
         tensor([8, 9]),
         tensor([10]))

torch.arange(12).chunk(6)
        (tensor([0, 1]),
         tensor([2, 3]),
         tensor([4, 5]),
         tensor([6, 7]),
         tensor([8, 9]),
         tensor([10, 11]))
torch.arange(13).chunk(6)
```

```
(tensor([0, 1, 2]),
 tensor([3, 4, 5]),
 tensor([6, 7, 8]),
 tensor([ 9, 10, 11]),
 tensor([12]))
```

A tensor can be split between multiple chunks. Those small chunks can be created along dim rows and dim columns. The following example shows a sample tensor of size (4,4). The chunk is created using the third argument in the function, as 0 or 1.

```
a = torch.randn(4, 4)
print(a)

torch.chunk(a,2)
tensor([[-0.5899, -1.3432, -1.0576, -0.1696],
        [ 0.2623, -0.1585,  1.0178, -0.2216],
        [-1.1716, -1.2771,  0.8073, -0.7717],
        [ 0.1768,  0.6423, -0.3200, -0.0480]])
(tensor([[-0.5899, -1.3432, -1.0576, -0.1696],
        [ 0.2623, -0.1585,  1.0178, -0.2216]]),
 tensor([[-1.1716, -1.2771,  0.8073, -0.7717],
        [ 0.1768,  0.6423, -0.3200, -0.0480]]))

torch.chunk(a,2,0)
(tensor([[-0.5899, -1.3432, -1.0576, -0.1696], [ 0.2623, -0.1585, 1.0178,
-0.2216]]), tensor([[-1.1716, -1.2771, 0.8073, -0.7717], [ 0.1768, 0.6423,
-0.3200, -0.0480]]))

torch.chunk(a,2,1)
(tensor([[-0.5899, -1.3432], [ 0.2623, -0.1585], [-1.1716, -1.2771], [
0.1768, 0.6423]]), tensor([[-1.0576, -0.1696], [ 1.0178, -0.2216], [
0.8073, -0.7717], [-0.3200, -0.0480]]))

torch.Tensor([[11,12],[23,24]])
tensor([[11., 12.], [23., 24.]])
```

The gather function collects elements from a tensor and places them in another tensor using an index argument. The index position is determined by the LongTensor function in PyTorch.

```
torch.gather(torch.Tensor([[11,12],[23,24]]), 1,
            torch.LongTensor([[0,0],[1,0]]))
tensor([[11., 11.], [24., 23.]])
```

```
torch.LongTensor([[0,0],[1,0]])
#the 1D tensor containing the indices to index
tensor([[0, 0], [1, 0]])
```

The LongTensor function or the index select function can be used to fetch relevant values from a tensor. The following sample code shows two options: selection along rows and selection along columns. If the second argument is 0, it is for rows. If it is 1, then it is along the columns.

```
a = torch.randn(4, 4)
print(a)
tensor([[-0.9183, -2.3470,  1.5208, -0.1585],
        [-0.6741, -0.6297,  0.2581,  1.1954],
        [ 1.0443, -1.3408,  0.7863, -0.6056],
        [-0.6946, -0.5963,  0.1936, -2.0625]])
```

```
indices = torch.LongTensor([0, 2])
torch.index_select(a, 0, indices)
tensor([[-0.9183, -2.3470, 1.5208, -0.1585], [ 1.0443, -1.3408, 0.7863,
-0.6056]])
```

```
torch.index_select(a, 1, indices)
tensor([[-0.9183, 1.5208], [-0.6741, 0.2581], [ 1.0443, 0.7863], [-0.6946,
0.1936]])
```

It is a common practice to check non-missing values in a tensor. The objective is to identify non-zero elements in a large tensor.

```
#identify null input tensors using nonzero function
torch.nonzero(torch.tensor([10,00,23,0,0.0]))
tensor([[0], [2]]) torch.nonzero(torch.Tensor([10,00,23,0,0.0]))
tensor([[0], [2]])
```

Restructuring the input tensors into smaller tensors not only fastens the calculation process, but also helps in distributed computing. The split function splits a long tensor into smaller tensors.

```
# splitting the tensor into small chunks
torch.split(torch.tensor([12,21,34,32,45,54,56,65]),2)
(tensor([12, 21]), tensor([34, 32]), tensor([45, 54]), tensor([56, 65]))
```

```
# splitting the tensor into small chunks
torch.split(torch.tensor([12,21,34,32,45,54,56,65]),3)
(tensor([12, 21, 34]), tensor([32, 45, 54]), tensor([56, 65]))
```

```
torch.zeros(3,2,4)
tensor([[[0., 0., 0., 0.], [0., 0., 0., 0.]], [[0., 0., 0., 0.], [0., 0.,
0., 0.]], [[0., 0., 0., 0.], [0., 0., 0., 0.]]])
```

```
torch.zeros(3,2,4).size()
torch.Size([3, 2, 4])
```

Now let's have a look at examples of how the input tensor can be resized given the computational difficulty. The transpose function is primarily used to reshape tensors. There are two ways of writing the transpose function: .t and .transpose.

```
#how to reshape the tensors along a new dimension
x
tensor([[-1.5343, -1.3533, -0.8621, -1.1674, -0.1114], [ 0.2790, 0.0463,
1.5364, -0.1287, 0.6379], [-0.4542, 0.5196, 0.2335, -0.5135, -0.6602],
[-0.6930, 0.0541, -0.8463, -0.4498, -0.0280]])
```

```
x.t() #transpose is one option to change the shape of the tensor
tensor([[-1.5343, 0.2790, -0.4542, -0.6930], [-1.3533, 0.0463, 0.5196,
0.0541], [-0.8621, 1.5364, 0.2335, -0.8463], [-1.1674, -0.1287, -0.5135,
-0.4498], [-0.1114, 0.6379, -0.6602, -0.0280]])
```

```
# transpose partially based on rows and columns
x.transpose(1,0)
tensor([[-1.5343, 0.2790, -0.4542, -0.6930], [-1.3533, 0.0463, 0.5196,
0.0541], [-0.8621, 1.5364, 0.2335, -0.8463], [-1.1674, -0.1287, -0.5135,
-0.4498], [-0.1114, 0.6379, -0.6602, -0.0280]])
```

```
# how to remove a dimension from a tensor
x
tensor([[-1.5343, -1.3533, -0.8621, -1.1674, -0.1114], [ 0.2790, 0.0463,
1.5364, -0.1287, 0.6379], [-0.4542, 0.5196, 0.2335, -0.5135, -0.6602],
[-0.6930, 0.0541, -0.8463, -0.4498, -0.0280]])
```

The unbind function removes a dimension from a tensor. To remove the dimension row, the 0 value needs to be passed. To remove a column, the 1 value needs to be passed.

```
torch.unbind(x,1) #dim=1 removing a column
(tensor([-1.5343, 0.2790, -0.4542, -0.6930]), tensor([-1.3533,
0.0463, 0.5196, 0.0541]), tensor([-0.8621, 1.5364, 0.2335, -0.8463]),
tensor([-1.1674, -0.1287, -0.5135, -0.4498]), tensor([-0.1114, 0.6379,
-0.6602, -0.0280]))
```

```
torch.unbind(x) #dim=0 removing a row
(tensor([-1.5343, -1.3533, -0.8621,  1.1674, -0.1114]), tensor([ 0.2790,
0.0463, 1.5364, -0.1287, 0.6379]), tensor([-0.4542, 0.5196, 0.2335,
-0.5135, -0.6602]), tensor([-0.6930, 0.0541, -0.8463, -0.4498, -0.0280]))
```

```
x
tensor([[-1.5343, -1.3533, -0.8621, -1.1674, -0.1114], [ 0.2790, 0.0463,
1.5364, -0.1287, 0.6379], [-0.4542, 0.5196, 0.2335, -0.5135, -0.6602],
[-0.6930, 0.0541, -0.8463, -0.4498, -0.0280]])
```

```
#how to compute the basic mathematrical functions
torch.abs(torch.FloatTensor([-10, -23, 3.000]))
tensor([10., 23., 3.])
```

Mathematical functions are the backbone of implementing any algorithm in PyTorch, so let's go through functions that help perform arithmetic-based operations. A scalar is a single value, and a tensor 1D is a row, like NumPy. The scalar multiplication and addition with a 1D tensor are done using the add and mul functions. The following script shows scalar addition and multiplication with a tensor:

```
#adding value to the existing tensor, scalar addition
torch.add(x,20)
```

```
tensor([[18.4657, 18.6467, 19.1379, 18.8326, 19.8886], [20.2790, 20.0463,
21.5364, 19.8713, 20.6379], [19.5458, 20.5196, 20.2335, 19.4865, 19.3398],
[19.3070, 20.0541, 19.1537, 19.5502, 19.9720]])

x
tensor([[-1.5343, -1.3533, -0.8621, -1.1674, -0.1114], [ 0.2790, 0.0463,
1.5364, -0.1287, 0.6379], [-0.4542, 0.5196, 0.2335, -0.5135, -0.6602],
[-0.6930, 0.0541, -0.8463, -0.4498, -0.0280]])

# scalar multiplication
torch.mul(x,2)
tensor([[-3.0686, -2.7065, -1.7242, -2.3349, -0.2227], [ 0.5581, 0.0926,
3.0727, -0.2575, 1.2757], [-0.9084, 1.0392, 0.4670, -1.0270, -1.3203],
[-1.3859, 0.1082, -1.6926, -0.8995, -0.0560]])

x
tensor([[-1.5343, -1.3533, -0.8621, -1.1674, -0.1114], [ 0.2790, 0.0463,
1.5364, -0.1287, 0.6379], [-0.4542, 0.5196, 0.2335, -0.5135, -0.6602],
[-0.6930, 0.0541, -0.8463, -0.4498, -0.0280]])
```

Combined mathematical operations, such as expressing linear equations as tensor operations, can be done using the following sample script. Here you express the outcome y object as a linear combination of beta values times the independent x object, plus the constant term.

```
# how do we represent the equation in the form of a tensor
# y = intercept + (beta * x)
intercept = torch.randn(1)
intercept
tensor([-1.1444])

x = torch.randn(2, 2)
x
tensor([[ 1.3517, -0.3991], [-0.4170, -0.1862]])

beta = 0.7456
beta
0.7456
```

Output = Constant + (beta * Independent)

```
torch.mul(x,beta)
tensor([[ 1.0078, -0.2976], [-0.3109, -0.1388]])
```

```
torch.add(x,beta,intercept)
tensor([[ 0.4984, -1.2524], [-1.2703, -1.0395]])
```

```
torch.mul(intercept,x)
tensor([[-1.5469, 0.4568], [ 0.4773, 0.2131]])
```

```
torch.mul(x,beta)
tensor([[ 1.0078, -0.2976], [-0.3109, -0.1388]])
```

```
## y = intercept + (beta * x)
torch.add(torch.mul(intercept,x),torch.mul(x,beta)) # tensor y
tensor([[-0.5391, 0.1592], [ 0.1663, 0.0743]])
```

Like Numpy operations, the element-wise matrix multiplication also can be done using tensors. There are two different ways of doing matrix multiplication: element-wise and combined together.

```
tensor
tensor([[[ 4, 64, 5, 4], [ 10, 20, 30, 110], [ 45, 34, 67, 40], [ 56, 67,
89, 90]]])
```

```
# Element-wise matrix mutlication
tensor * tensor
tensor([[[ 16, 4096, 25, 16], [ 100, 400, 900, 12100], [ 2025, 1156, 4489,
1600], [ 3136, 4489, 7921, 8100]]])
```

```
torch.matmul(tensor, tensor)
tensor([[[ 1105, 1974, 2631, 7616], [ 7750, 9430, 12450, 13340], [ 5775,
8518, 9294, 10200], [ 9939, 13980, 16263, 19254]]])
```

```
tensor @ tensor
tensor([[[ 1105, 1974, 2631, 7616], [ 7750, 9430, 12450, 13340], [ 5775,
8518, 9294, 10200], [ 9939, 13980, 16263, 19254]]])
```

Like NumPy operations, the tensor values must be rounded up by using either the ceiling or the flooring function, which is done using the following syntax:

```
# how to round up tensor values
torch.manual_seed(1234)
torch.randn(5,5)
tensor([[-0.1117, -0.4966, 0.1631, -0.8817, 0.0539], [ 0.6684, -0.0597,
-0.4675, -0.2153, -0.7141], [-1.0831, -0.5547, 0.9717, -0.5150, 1.4255],
[ 0.7987, -1.4949, 1.4778, -0.1696, -0.9919], [-1.4569, 0.2563, -0.4030,
0.4195, 0.9380]])
```

```
torch.manual_seed(1234)
torch.ceil(torch.randn(5,5))
tensor([[-0., -0., 1., -0., 1.], [ 1., -0., -0., -0., -0.], [-1., -0., 1.,
-0., 2.], [ 1., -1., 2., -0., -0.], [-1., 1., -0., 1., 1.]])
```

```
torch.manual_seed(1234)
torch.floor(torch.randn(5,5))
tensor([[-1., -1., 0., -1., 0.], [ 0., -1., -1., -1., -1.], [-2., -1., 0.,
-1., 1.], [ 0., -2., 1., -1., -1.], [-2., 0., -1., 0., 0.]])
```

Limiting the values of any tensor within a certain range can be done using the minimum and maximum argument and using the clamp function. The same function can apply minimum and maximum in parallel or any one of them to any tensor, be it 1D or 2D; 1D is the far simpler version. The following example shows the implementation in a 2D scenario:

```
# truncate the values in a range say 0,1
torch.manual_seed(1234)
torch.clamp(torch.floor(torch.randn(5,5)), min=-0.3, max=0.4)
tensor([[-0.3000, -0.3000, 0.0000, -0.3000, 0.0000], [ 0.0000, -0.3000,
-0.3000, -0.3000, -0.3000], [-0.3000, -0.3000, 0.0000, -0.3000, 0.4000],
[ 0.0000, -0.3000, 0.4000, -0.3000, -0.3000], [-0.3000, 0.0000, -0.3000,
0.0000, 0.0000]])
```

```
#truncate with only lower limit
torch.manual_seed(1234)
torch.clamp(torch.floor(torch.randn(5,5)), min=-0.3)
tensor([[-0.3000, -0.3000, 0.0000, -0.3000, 0.0000], [ 0.0000, -0.3000,
-0.3000, -0.3000, -0.3000], [-0.3000, -0.3000, 0.0000, -0.3000, 1.0000],
```

```
[ 0.0000, -0.3000, 1.0000, -0.3000, -0.3000], [-0.3000, 0.0000, -0.3000,
0.0000, 0.0000]])
```

```
#truncate with only upper limit
torch.manual_seed(1234)
torch.clamp(torch.floor(torch.randn(5,5)), max=0.3)
tensor([[-1.0000, -1.0000, 0.0000, -1.0000, 0.0000], [ 0.0000, -1.0000,
-1.0000, -1.0000, -1.0000], [-2.0000, -1.0000, 0.0000, -1.0000, 0.3000],
[ 0.0000, -2.0000, 0.3000, -1.0000, -1.0000], [-2.0000, 0.0000, -1.0000,
0.0000, 0.0000]])
```

How do you get the exponential of a tensor? How do you get the fractional portion of the tensor if it has decimal places and is defined as a floating data type?

```
#scalar division
torch.div(x,0.10)
tensor([[13.5168, -3.9914], [-4.1705, -1.8621]])
```

```
#compute the exponential of a tensor
torch.exp(x)
tensor([[3.8639, 0.6709], [0.6590, 0.8301]])
```

```
np.exp(x)
tensor([[3.8639, 0.6709], [0.6590, 0.8301]])
```

```
#how to get the fractional portion of each tensor
torch.add(x,10)
tensor([[11.3517, 9.6009], [ 9.5830, 9.8138]])
```

```
torch.frac(torch.add(x,10))
tensor([[0.3517, 0.6009], [0.5830, 0.8138]])
```

The following syntax explains the logarithmic values in a tensor. The values with a negative sign are converted to nan. The power function computes the exponential of any value in a tensor.

```
# compute the log of the values in a tensor
x
tensor([[ 1.3517, -0.3991], [-0.4170, -0.1862]])
```

```
torch.log(x) #log of negatives are nan
tensor([[0.3013, nan], [ nan, nan]])

# to rectify the negative values do a power tranforamtion
torch.pow(x,2)
tensor([[1.8270, 0.1593], [0.1739, 0.0347]])

# rounding up similar to numpy
x
tensor([[ 1.3517, -0.3991], [-0.4170, -0.1862]])

np.round(x)
tensor([[1., -0.], [-0., -0.]])

torch.round(x)
tensor([[1., -0.], [-0., -0.]])
```

To compute the transformation functions (i.e., sigmoid, hyperbolic tangent, and radial basis function, which are the most commonly used transfer functions in deep learning), you must construct the tensors. The following sample script shows how to create a sigmoid function and apply it on a tensor:

```
# how to compute the sigmoid of the input tensor
x
tensor([[ 1.3517, -0.3991], [-0.4170, -0.1862]])

torch.sigmoid(x)
tensor([[0.7944, 0.4015], [0.3972, 0.4536]])

# finding the square root of the values
x
tensor([[ 1.3517, -0.3991], [-0.4170, -0.1862]])

torch.sqrt(x)
tensor([[1.1626, nan], [ nan, nan]])

# Create a tensor
x = torch.arange(10, 10000, 150)
x
```

```
tensor([ 10, 160, 310, 460, 610, 760, 910, 1060, 1210, 1360, 1510, 1660,
1810, 1960, 2110, 2260, 2410, 2560, 2710, 2860, 3010, 3160, 3310, 3460,
3610, 3760, 3910, 4060, 4210, 4360, 4510, 4660, 4810, 4960, 5110, 5260,
5410, 5560, 5710, 5860, 6010, 6160, 6310, 6460, 6610, 6760, 6910, 7060,
7210, 7360, 7510, 7660, 7810, 7960, 8110, 8260, 8410, 8560, 8710, 8860,
9010, 9160, 9310, 9460, 9610, 9760, 9910])
```

```
print(f"Minimum: {x.min()}")
print(f"Maximum: {x.max()}")
# print(f"Mean: {x.mean()}") # this will error
print(f"Mean: {x.type(torch.float32).mean()}") # won't work without float
datatype
print(f"Sum: {x.sum()}")
Minimum: 10
Maximum: 9910
Mean: 4960.0
Sum: 332320
```

```
torch.argmax(x),torch.argmin(x)
(tensor(66), tensor(0))
```

```
torch.max(x),torch.min(x)
(tensor(9910), tensor(10))
```

```
# how to change data type
y = torch.tensor([[39,339.63],
[36,667.20],
[33,978.07],
[31,897.13],
[29,178.19],
[26,442.25],
[24,314.22],
[21,547.88],
[18,764.25],
[16,588.23],
[13,773.61]],dtype=torch.float32)
```

```
# Create a float16 tensor
tensor_float16 = y.type(torch.float16)
tensor_float16
tensor([[ 39.0000, 339.7500], [ 36.0000, 667.0000], [ 33.0000, 978.0000],
[ 31.0000, 897.0000], [ 29.0000, 178.2500], [ 26.0000, 442.2500], [
24.0000, 314.2500], [ 21.0000, 548.0000], [ 18.0000, 764.0000], [ 16.0000,
588.0000], [ 13.0000, 773.5000]], dtype=torch.float16)
```

```
# Create a int8 tensor
tensor_int8 = y.type(torch.int8)
tensor_int8
tensor([[ 39, 83], [ 36, -101], [ 33, -46], [ 31, -127], [ 29, -78],
[ 26, -70], [ 24, 58], [ 21, 35], [ 18, -4], [ 16, 76], [ 13, 5]],
dtype=torch.int8)
```

```
# Change view (keeps same data as original but changes view)
y.view(2,11)
tensor([[ 39.0000, 339.6300, 36.0000, 667.2000, 33.0000, 978.0700,
31.0000, 897.1300, 29.0000, 178.1900, 26.0000], [442.2500, 24.0000,
314.2200, 21.0000, 547.8800, 18.0000, 764.2500, 16.0000, 588.2300, 13.0000,
773.6100]])
```

```
# stacking and unstacking of tensors
A = torch.arange(10,50,5)
B = torch.arange(20,60,5)
torch.stack([A,B],dim=0)
tensor([[10, 15, 20, 25, 30, 35, 40, 45], [20, 25, 30, 35, 40, 45,
50, 55]])
```

```
torch.stack([A,B],dim=1)
tensor([[10, 20], [15, 25], [20, 30], [25, 35], [30, 40], [35, 45], [40,
50], [45, 55]])
```

```
torch.stack([A,B])
tensor([[10, 15, 20, 25, 30, 35, 40, 45], [20, 25, 30, 35, 40, 45,
50, 55]])
```

```
# indexing of tensors
```

```
y = torch.stack([A,B,A,B,A,B,A,B])
y
tensor([[10, 15, 20, 25, 30, 35, 40, 45], [20, 25, 30, 35, 40, 45, 50, 55],
[10, 15, 20, 25, 30, 35, 40, 45], [20, 25, 30, 35, 40, 45, 50, 55], [10,
15, 20, 25, 30, 35, 40, 45], [20, 25, 30, 35, 40, 45, 50, 55], [10, 15, 20,
25, 30, 35, 40, 45], [20, 25, 30, 35, 40, 45, 50, 55]])

# Get all values of 0th dimension and the 1st index of 1st dimension
y[:, 1]
tensor([15, 25, 15, 25, 15, 25, 15, 25])

D = torch.tensor([[[12,13,14],
                   [15,16,17],
                   [18,19,20]]])
# Get all values of 0th & 1st dimensions but only index 1 of 2nd dimension
D[:, :, 1]
tensor([[13, 16, 19]])

# Get all values of the 0 dimension but only the 1 index value of the 1st
and 2nd dimension
D[:, 1, 1]
tensor([16])

# Get index 0 of 0th and 1st dimension and all values of 2nd dimension
D[0, 0, :] # same as D[0][0]
tensor([12, 13, 14])

D[0][0]
tensor([12, 13, 14])

# Check for GPU
import torch
torch.cuda.is_available()
False

# Set device type
device = "cuda" if torch.cuda.is_available() else "cpu"
device
cpu
```

Following the CUDA semantics, PyTorch can be configured for a GPU, which is given at this link: https://pytorch.org/docs/stable/notes/cuda.html.

```
# Count number of devices
torch.cuda.device_count()

    0

# x = torch.randn(2, 2, device='cpu') #on cpu
# x = torch.randn(2, 2, device='gpu') #on gpu

# x = torch.randn(2, 2, device=device)
```

The syntax of generating random numbers are device agnostic, so it works on both CPU and GPU environments.

```
# flatten tensor like numpy
D.flatten()
tensor([12, 13, 14, 15, 16, 17, 18, 19, 20])

# Concatenate along rows
cat_rows = torch.cat((A, B), dim=0)
cat_rows
tensor([10, 15, 20, 25, 30, 35, 40, 45, 20, 25, 30, 35, 40, 45, 50, 55])

cat_cols = torch.cat((A.reshape(2,4), B.reshape(2,4)), dim=1)
cat_cols
tensor([[10, 15, 20, 25, 20, 25, 30, 35], [30, 35, 40, 45, 40, 45,
50, 55]])
```

Conclusion

This chapter is a refresher for people who have prior experience in PyTorch and Python. It is a basic building block for people who are new to the PyTorch framework. Before starting the advanced topics, it is important to be familiar with the terminology and basic syntaxes. The next chapter is on using PyTorch to implement probabilistic models, which includes the creation of random variables, the application of statistical distributions, and making statistical inferences.

## CHAPTER 2

Probability Distributions Using PyTorch

Probability and random variables are an integral part of computation in a graph-computing platform like PyTorch. You must have an understanding of probability and associated concepts. This chapter covers probability distributions and implementation using PyTorch and interpreting results from tests.

In probability and statistics, a random variable is also known as a *stochastic variable*, whose outcome is dependent on a purely stochastic phenomenon or random phenomenon. There are different types of probability distributions, including normal distribution, binomial distribution, multinomial distribution, and Bernoulli distribution. Each statistical distribution has its own properties.

The `torch.distributions` module contains probability distributions and sampling functions. Each distribution type has its own importance in a computational graph. The `distributions` module contains binomial, Bernoulli, beta, categorical, exponential, normal, and Poisson distributions.

Recipe 2-1. Sampling Tensors
Problem

Weight initialization is an important task in training a neural network and any kind of deep learning model, such as a convolutional neural network (CNN), a deep neural network (DNN), and a recurrent neural network (RNN). The question is always how to initialize the weights.

© Pradeepta Mishra 2023
P. Mishra, *PyTorch Recipes*, https://doi.org/10.1007/978-1-4842-8925-9_2

Solution

Weight initialization can be done using various methods, including random weight initialization. Weight initialization can be done based on a distribution including uniform distribution, Bernoulli distribution, multinomial distribution, and normal distribution. How to do it using PyTorch is explained next.

How It Works

To execute a neural network, a set of initial weights needs to be passed to the backpropagation layer to compute the loss function (and hence, the accuracy can be calculated). The selection of a method depends on the data type, the task, and the optimization required for the model. Here you are going to look at all types of approaches to initialize weights.

If the use case requires reproducing the same set of results to maintain consistency, then a manual seed needs to be set.

```
import torch
print(torch.cuda.is_available())
False
```

```
# CUDA is an API developed by NVIDIA to interface GPU
x = torch.randn(10)
print(x.device)
cpu
```

```
# how to perform random sampling of the tensors
torch.manual_seed(1234)
```

```
torch.manual_seed(1234)
torch.randn(4,4)
tensor([[-0.1117, -0.4966, 0.1631, -0.8817], [ 0.0539, 0.6684, -0.0597,
-0.4675], [-0.2153, 0.8840, -0.7584, -0.3689], [-0.3424, -1.4020, 0.3206,
-1.0219]])
```

The seed value can be customized. The random number is generated purely by chance. Random numbers can also be generated from a statistical distribution. The probability density function of the *continuous uniform distribution* is defined by the following formula:

$$f(x) = \begin{cases} \dfrac{1}{b-a} & \text{for } a \le x \le b, \\ 0 & \text{for } x\langle a \ \text{ or } \ x\rangle b \end{cases}$$

The function of x has two points, a and b, in which a is the starting point and b is the end. In a continuous uniform distribution, each number has an equal chance of being selected. In the following example, the start is 0 and the end is 1; between those two digits, all 16 elements are selected randomly:

```
#generate random numbers from a statistical distribution
torch.Tensor(4, 4).uniform_(0, 1) #random number from uniform distribution
tensor([[0.2837, 0.6567, 0.2388, 0.7313], [0.6012, 0.3043, 0.2548, 0.6294],
[0.9665, 0.7399, 0.4517, 0.4757], [0.7842, 0.1525, 0.6662, 0.3343]])
```

In statistics, the *Bernoulli distribution* is considered as the discrete probability distribution, which has two possible outcomes. If the event happens, then the value is 1, and if the event does not happen, then the value is 0.

For *discrete probability distribution,* you calculate probability mass function instead of probability density function. The probability mass function looks like this:

$$\begin{cases} q = (1-p) & \text{for} k = 0 \\ p & \text{for} k = 1 \end{cases}$$

From the Bernoulli distribution, you create sample tensors by considering the uniform distribution of size 4 and 4 in a matrix format, as follows:

```
#now apply the distribution assuming the input values from the
#tensor are probabilities
torch.bernoulli(torch.Tensor(4, 4).uniform_(0, 1))
tensor([[0., 0., 0., 0.], [1., 0., 1., 0.], [1., 0., 1., 1.], [0., 0.,
0., 0.]])
```

The generation of sample random values from a *multinomial distribution* is defined by the following script. In a multinomial distribution, you can choose with a replacement or without a replacement. By default, the multinomial function picks up without a replacement and returns the result as an index position for the tensors. If you need to run it with a replacement, then you need to specify that while sampling.

```
#how to perform sampling from a multinomial distribution
torch.Tensor([10, 10, 13, 10,34,45,65,67,87,89,87,34])
tensor([10., 10., 13., 10., 34., 45., 65., 67., 87., 89., 87., 34.])

torch.multinomial(torch.tensor([10., 10., 13., 10.,
                                34., 45., 65., 67.,
                                87., 89., 87., 34.]),
                3)
tensor([4, 5, 7])
```

Sampling from multinomial distribution with a replacement returns the tensors' index values.

```
torch.multinomial(torch.tensor([10., 10., 13., 10.,
                                34., 45., 65., 67.,
                                87., 89., 87., 34.]),
                5, replacement=True)
tensor([10, 5, 9, 10, 5])
```

The weight initialization from the normal distribution is a method that is used in fitting a neural network or deep neural network and CNN and RNN. Let's have a look at the process of creating a set of random weights generated from a normal distribution.

```
#generate random numbers from the normal distribution
torch.normal(mean=torch.arange(1., 11.),
           std=torch.arange(1, 0, -0.1))
tensor([1.5236, 2.2441, 2.7375, 3.9521, 5.4380, 5.5158, 8.2489, 8.1645,
9.0575, 9.8627])

torch.normal(mean=0.5,
           std=torch.arange(1., 6.))
tensor([ 1.1144, 0.0361, 1.2766, -1.3999, -0.1648])

torch.normal(mean=0.5,
           std=torch.arange(0.2,0.6))
tensor([-0.0844])
```

```
#computing the descriptive statistics: mean
torch.mean(torch.tensor([10., 10., 13., 10., 34.,
                         45., 65., 67., 87., 89., 87., 34.]))
tensor(45.9167)

# mean across rows and across columns
d = torch.randn(4, 5)
d
tensor([[-1.6406, 0.9295, 1.2907, 0.2612, 0.9711], [ 0.3551, 0.8562,
-0.3635, -0.1552, -1.2282], [ 1.2445, 1.1750, -0.2217, -2.0901, -1.2658],
[-1.8761, -0.6066, 0.7470, 0.4811, 0.6234]])

torch.mean(d,dim=0)
tensor([-0.4793, 0.5885, 0.3631, -0.3757, -0.2249])

torch.mean(d,dim=1)
tensor([ 0.3624, -0.1071, -0.2316, -0.1262])

#compute median
torch.median(d,dim=0)
torch.return_types.median( values=tensor([-1.6406, 0.8562, -0.2217,
-0.1552, -1.2282]), indices=tensor([0, 1, 2, 1, 1]))

torch.median(d,dim=1)
torch.return_types.median( values=tensor([ 0.9295, -0.1552, -0.2217,
0.4811]), indices=tensor([1, 3, 2, 3]))
```

Recipe 2-2. Variable Tensors

Problem

What is a variable in PyTorch and how is it defined? What is a random variable in
PyTorch?

Solution

In PyTorch, algorithms are represented as a computational graph. A variable is considered as a representation around the tensor object, corresponding gradients, and a reference to the function from where it was created. For simplicity, gradients are considered as the slope of the function. The slope of the function can be computed by the derivative of the function with respect to the parameters that are present in the function. For example, in linear regression (Y = W*X + alpha), representation of the variable looks like the one shown in Figure 2-1.

Basically, a PyTorch variable is a node in a computational graph that stores data and gradients. When training a neural network model, after each iteration, you must compute the gradient of the loss function with respect to the parameters of the model, such as weights and biases. After that, you usually update the weights using the gradient descent algorithm. Figure 2-1 explains how the linear regression equation is deployed under the hood using a neural network model in the PyTorch framework.

In a computational graph structure, the sequencing and ordering of tasks is very important. The one-dimensional tensors are X, Y, W, and alpha in Figure 2-1. The direction of the arrows changes when you implement backpropagation to update the weights to match with Y, so that the error or loss function between Y and predicted Y can be minimized.

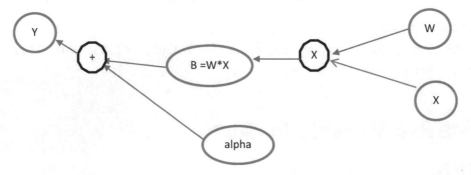

Figure 2-1. _A sample computational graph of a PyTorch implementation_

How It Works

An example of how a variable is used to create a computational graph is displayed in the following script. There are three variable objects around tensors— x1, x2, and x3—with random points generated from $a = 12$ and $b = 23$. The graph computation involves only multiplication and addition, and the final result with the gradient is shown.

The partial derivative of the loss function with respect to the weights and biases in a neural network model is achieved in PyTorch using the Autograd module. Variables are specifically designed to hold the changed values while running a backpropagation in a neural network model when the parameters of the model change. The variable type is just a wrapper around the tensor. It has three properties: data, grad, and function.

```
from torch.autograd import Variable
Variable(torch.ones(2,2),requires_grad=True)
tensor([[1., 1.], [1., 1.]], requires_grad=True)

a, b = 12,23
x1 = Variable(torch.randn(a,b),
            requires_grad=True)
x2 = Variable(torch.randn(a,b),
            requires_grad=True)
x3 =Variable(torch.randn(a,b),
            requires_grad=True)
c = x1 * x2
d = a + x3
e = torch.sum(d)

e.backward()

print(e)
tensor(3278.1235, grad_fn=<SumBackward0>)

x1.data
tensor([[-4.9545e-02, 6.2245e-01, 1.6573e-01, 3.1583e-01, 2.4915e-01,
-4.9784e-01, 2.9079e+00, 1.6201e+00, -6.4459e-01, -1.9885e-02, 1.6222e+00,
1.4239e+00, 9.0691e-01, 7.6310e-02, 1.1225e+00, -1.2433e+00, -6.7258e-01,
8.8433e-01, -6.6589e-01, -7.3347e-01, -2.7599e-01, 5.5485e-01,
-1.9303e+00],................

x2.data
tensor([[-7.5597e-01, -1.1689e+00, -9.3890e-01, 8.8566e-01, 1.3764e+00,
-7.8276e-01, 2.2200e-01, 7.3758e-02, -6.9147e-01, -5.1308e-01, 1.1427e+00,
-1.0126e+00, 1.1602e-01, -1.0350e+00, 1.0803e+00, -7.9977e-01,
```

```
-9.1219e-02, 5.0242e-01, -4.5173e-01, -4.8067e-01, 5.9066e-01, 1.6343e-01,
-3.1368e-02],.............
```

```
x3.data
tensor([[ 0.2499, 0.2458, 0.1029, -0.6494, -0.3258, 0.8149, 0.4049, 0.2481,
0.4841, 0.3293, -1.2471, 0.2117, 1.4315, 0.0502, -0.3668, 0.8378, -0.7901,
0.0267, -0.3120, 2.4534, 0.7926, 0.2382, -0.5245].........
```

Recipe 2-3. Basic Statistics
Problem

How can you compute basic statistics, such as mean, median, mode, and so forth, from a Torch tensor?

Solution

Computation of basic statistics using PyTorch enables you to apply probability distributions and statistical tests to make inferences from data. Although the Torch functionality is like that of Numpy, Torch functions have GPU acceleration. Let's have a look at the functions to create basic statistics.

How It Works

The mean computation is simple to write for a 1D tensor; however, for a 2D tensor, an extra argument needs to be passed as a mean, median, or mode computation, across which the dimension needs to be specified.

```
#computing the descriptive statistics: mean
torch.mean(torch.tensor([10., 10., 13., 10., 34.,
                         45., 65., 67., 87., 89., 87., 34.]))
tensor(45.9167)
```

```
# mean across rows and across columns
d = torch.randn(4, 5)
d
```

```
tensor([[-1.6406, 0.9295, 1.2907, 0.2612, 0.9711], [ 0.3551, 0.8562,
-0.3635, -0.1552, -1.2282], [ 1.2445, 1.1750, -0.2217, -2.0901, -1.2658],
[-1.8761, -0.6066, 0.7470, 0.4811, 0.6234]])
```

```
torch.mean(d,dim=0)
tensor([-0.4793, 0.5885, 0.3631, -0.3757, -0.2249])
```

```
torch.mean(d,dim=1)
tensor([ 0.3624, -0.1071, -0.2316, -0.1262])
```

Median, mode, and standard deviation computation can be written in the same way.

```
#compute median
torch.median(d,dim=0)
torch.return_types.median( values=tensor([-1.6406, 0.8562, -0.2217,
-0.1552, -1.2282]), indices=tensor([0, 1, 2, 1, 1]))
```

```
torch.median(d,dim=1)
torch.return_types.median( values=tensor([ 0.9295, -0.1552, -0.2217,
0.4811]), indices=tensor([1, 3, 2, 3]))
```

```
# compute the mode
torch.mode(d)
torch.return_types.mode( values=tensor([-1.6406, -1.2282, -2.0901,
-1.8761]), indices=tensor([0, 4, 3, 0]))
```

```
torch.mode(d,dim=0)
torch.return_types.mode( values=tensor([-1.8761, -0.6066, -0.3635, -2.0901,
-1.2658]), indices=tensor([3, 3, 1, 2, 2]))
```

```
torch.mode(d,dim=1)
torch.return_types.mode( values=tensor([-1.6406, -1.2282, -2.0901,
-1.8761]), indices=tensor([0, 4, 3, 0]))
```

Standard deviation shows the deviation from the measures of central tendency, which indicates the consistency of the data/variable. It shows whether there is enough fluctuation in data or not.

```
#compute the standard deviation
torch.std(d)
tensor(1.0944)
```

```
torch.std(d,dim=0)
tensor([1.5240, 0.8083, 0.7911, 1.1730, 1.1889])

torch.std(d,dim=1)
tensor([1.1807, 0.7852, 1.4732, 1.1165])

#compute variance
torch.var(d)
tensor(1.1978)

torch.var(d,dim=0)
tensor([2.3224, 0.6534, 0.6259, 1.3758, 1.4134])

torch.var(d,dim=1)
tensor([1.3940, 0.6166, 2.1703, 1.2466])

# compute min and max
torch.min(d)
tensor(-2.0901)

torch.min(d,dim=0)
torch.return_types.min( values=tensor([-1.8761, -0.6066, -0.3635, -2.0901,
-1.2658]), indices=tensor([3, 3, 1, 2, 2]))

torch.min(d,dim=1)
torch.return_types.min( values=tensor([-1.6406, -1.2282, -2.0901,
-1.8761]), indices=tensor([0, 4, 3, 0]))

# sorting a tensor
torch.sort(d)
torch.return_types.sort( values=tensor([[-1.6406, 0.2612, 0.9295, 0.9711,
1.2907], [-1.2282, -0.3635, -0.1552, 0.3551, 0.8562], [-2.0901, -1.2658,
-0.2217, 1.1750, 1.2445], [-1.8761, -0.6066, 0.4811, 0.6234, 0.7470]]),
indices=tensor([[0, 3, 1, 4, 2], [4, 2, 3, 0, 1], [3, 4, 2, 1, 0], [0, 1,
3, 4, 2]]))

torch.sort(d,dim=0)
torch.return_types.sort( values=tensor([[-1.8761, -0.6066, -0.3635,
-2.0901, -1.2658], [-1.6406, 0.8562, -0.2217, -0.1552, -1.2282], [ 0.3551,
0.9295, 0.7470, 0.2612, 0.6234], [ 1.2445, 1.1750, 1.2907, 0.4811,
```

```
0.9711]]), indices=tensor([[3, 3, 1, 2, 2], [0, 1, 2, 1, 1], [1, 0, 3, 0,
3], [2, 2, 0, 3, 0]]))

torch.sort(d,dim=0,descending=True)
torch.return_types.sort( values=tensor([[ 1.2445, 1.1750, 1.2907,
0.4811, 0.9711], [ 0.3551, 0.9295, 0.7470, 0.2612, 0.6234], [-1.6406,
0.8562, -0.2217, -0.1552, -1.2282], [-1.8761, -0.6066, -0.3635, -2.0901,
-1.2658]]), indices=tensor([[2, 2, 0, 3, 0], [1, 0, 3, 0, 3], [0, 1, 2, 1,
1], [3, 3, 1, 2, 2]]))

torch.sort(d,dim=1,descending=True)
torch.return_types.sort( values=tensor([[ 1.2907, 0.9711, 0.9295, 0.2612,
-1.6406], [ 0.8562, 0.3551, -0.1552, -0.3635, -1.2282], [ 1.2445, 1.1750,
-0.2217, -1.2658, -2.0901], [ 0.7470, 0.6234, 0.4811, -0.6066, -1.8761]]),
indices=tensor([[2, 4, 1, 3, 0], [1, 0, 3, 2, 4], [0, 1, 2, 4, 3], [2, 4,
3, 1, 0]]))

from torch.autograd import Variable
Variable(torch.ones(2,2),requires_grad=True)
tensor([[1., 1.], [1., 1.]], requires_grad=True)

a, b = 12,23
x1 = Variable(torch.randn(a,b),
            requires_grad=True)
x2 = Variable(torch.randn(a,b),
            requires_grad=True)
x3 =Variable(torch.randn(a,b),
            requires_grad=True)

c = x1 * x2
d = a + x3
e = torch.sum(d)

e.backward()

print(e)

tensor(3278.1235, grad_fn=<SumBackward0>)
```

Recipe 2-4. Gradient Computation

Problem

How do you compute basic gradients from the sample tensors using PyTorch?

Solution

Consider a sample datase0074, where two variables (x and y) are present. With the initial weight given, can you computationally get the gradients after each iteration? Let's take a look at the example.

How It Works

x_data and y_data both are lists. Computing the gradient of the two data lists requires computation of a loss function, a forward pass, and training in a loop.

The forward function computes the matrix multiplication of the weight tensor with the input tensor.

```
from torch import FloatTensor
from torch.autograd import Variable

a = Variable(FloatTensor([5]))
weights = [Variable(FloatTensor([i]), requires_grad=True) for i in (12, 53,
91, 73)]
w1, w2, w3, w4 = weights
b = w1 * a
c = w2 * a
d = w3 * b + w4 * c
Loss = (10 - d)
Loss.backward()

for index, weight in enumerate(weights, start=1):
    gradient, *_ = weight.grad.data
    print(f"Gradient of w{index} w.r.t to Loss: {gradient}")
Gradient of w1 w.r.t to Loss: -455.0
Gradient of w2 w.r.t to Loss: -365.0
```

Gradient of w3 w.r.t to Loss: -60.0
Gradient of w4 w.r.t to Loss: -265.0

```
# Using forward pass
def forward(x):
    return x * w

import torch
from torch.autograd import Variable

x_data = [11.0, 22.0, 33.0]
y_data = [21.0, 14.0, 64.0]

w = Variable(torch.Tensor([1.0]),  requires_grad=True)  # Any random value


# Before training
print("predict (before training)",  4, forward(4).data[0])

# define the Loss function
def loss(x, y):
    y_pred = forward(x)
    return (y_pred - y) * (y_pred - y)

# Run the Training loop
for epoch in range(10):
    for x_val, y_val in zip(x_data, y_data):
        l = loss(x_val, y_val)
        l.backward()
        print("\tgrad: ", x_val, y_val, w.grad.data[0])
        w.data = w.data - 0.01 * w.grad.data

        # Manually set the gradients to zero after updating weights
        w.grad.data.zero_()

    print("progress:", epoch, l.data[0])
grad:  11.0 21.0 tensor(-220.)
    grad:  22.0 14.0 tensor(2481.6001)
    grad:  33.0 64.0 tensor(-51303.6484)
progress: 0 tensor(604238.8125)...................................................
```

```
# After training
print("predict (after training)",  4, forward(4).data[0])
predict (after training) 4 tensor(-9.2687e+24)
```

The following program shows how to compute the gradients from a loss function using the variable method on the tensor:

```
a = Variable(FloatTensor([5]))
weights = [Variable(FloatTensor([i]), requires_grad=True) for i in (12, 53,
91, 73)]
w1, w2, w3, w4 = weights
b = w1 * a
c = w2 * a
d = w3 * b + w4 * c
Loss = (10 - d)
Loss.backward()
```

Recipe 2-5. Tensor Operations
Problem

How do you compute or perform operations based on variables such as matrix multiplication?

Solution

Tensors are wrapped within the variable, which has three properties: `grad`, `volatile`, and `gradient`.

How It Works

Let's create a variable and extract the properties of the variable. This is required because weight update process requires gradient computation. By using the `mm` module, you can perform matrix multiplication.

```
x = Variable(torch.Tensor(4, 4).uniform_(-4, 5))
y = Variable(torch.Tensor(4, 4).uniform_(-3, 2))
```

```
# matrix multiplication
z = torch.mm(x, y)
print(z.size())
torch.Size([4, 4])
```

The following program shows the properties of the variable, which is a wrapper around the tensor:

```
z = Variable(torch.Tensor(4, 4).uniform_(-5, 5))
print(z)
tensor([[-0.3071, -3.6691, -2.8417, -1.1818],
        [-1.4654, -0.4344, -2.0130, -2.3842],
        [ 1.3962,  1.4962, -2.0996,  1.8881],
        [-1.9797,  0.2337, -1.0308,  0.1266]])
```

```
print('Requires Gradient : %s ' % (z.requires_grad))
print('Volatile : %s ' % (z.volatile))
print('Gradient : %s ' % (z.grad))
print(z.data)
Requires Gradient : False
Volatile : False
Gradient : None
tensor([[-0.3071, -3.6691, -2.8417, -1.1818],
        [-1.4654, -0.4344, -2.0130, -2.3842],
        [ 1.3962,  1.4962, -2.0996,  1.8881],
        [-1.9797,  0.2337, -1.0308,  0.1266]])
```

Recipe 2-6. Tensor Operations

Problem

How do you compute or perform operations based on variables such as matrix-vector computation and matrix-matrix and vector-vector calculation?

Solution

One of the necessary conditions for the success of matrix-based operations is that the length of the tensor needs to match or be compatible for the execution of algebraic expressions.

How It Works

The tensor definition of a scalar is just one number. A 1D tensor is a vector, and a 2D tensor is a matrix. When it extends to an n dimensional level, it can be generalized to only tensors. When performing algebraic computations in PyTorch, the dimension of a matrix and a vector or scalar should be compatible.

```
#tensor operations
mat1 = torch.FloatTensor(4,4).uniform_(0,1)
mat1
tensor([[0.9002, 0.9188, 0.1386, 0.3701], [0.1947, 0.2268, 0.9587, 0.2615],
[0.7256, 0.7673, 0.5667, 0.1863], [0.4642, 0.4016, 0.9981, 0.8452]])

mat2 = torch.FloatTensor(4,4).uniform_(0,1)
mat2
tensor([[0.4962, 0.4947, 0.8344, 0.6721], [0.1182, 0.5997, 0.8990, 0.8252],
[0.1466, 0.1093, 0.8135, 0.9047], [0.2486, 0.1873, 0.6159, 0.2471]])

vec1 = torch.FloatTensor(4).uniform_(0,1)
vec1
tensor([0.7582, 0.6879, 0.8949, 0.3995])

# scalar addition
mat1 + 10.5
tensor([[11.4002, 11.4188, 10.6386, 10.8701], [10.6947, 10.7268, 11.4587,
10.7615], [11.2256, 11.2673, 11.0667, 10.6863], [10.9642, 10.9016, 11.4981,
11.3452]])

# scalar subtraction
mat2 - 0.20
```

```
tensor([[ 0.2962, 0.2947, 0.6344, 0.4721], [-0.0818, 0.3997, 0.6990,
0.6252], [-0.0534, -0.0907, 0.6135, 0.7047], [ 0.0486, -0.0127, 0.4159,
0.0471]])
```

```
# vector and matrix addition
mat1 + vec1
tensor([[1.6584, 1.6067, 1.0335, 0.7695], [0.9530, 0.9147, 1.8537, 0.6610],
[1.4839, 1.4553, 1.4616, 0.5858], [1.2224, 1.0895, 1.8931, 1.2446]])
```

```
mat2 + vec1
tensor([[1.2544, 1.1826, 1.7293, 1.0716], [0.8764, 1.2876, 1.7939, 1.2247],
[0.9049, 0.7972, 1.7084, 1.3042], [1.0068, 0.8752, 1.5108, 0.6466]])
```

If the mat1 and the mat2 dimensions are different, they are not compatible for matrix addition or multiplication. If the dimension remains the same, you can multiply them. In the following script, the matrix addition throws an error when you multiply similar dimensions—mat1 with mat1. You get relevant results.

```
# matrix-matrix addition
mat1 + mat2
tensor([[1.3963, 1.4135, 0.9730, 1.0422], [0.3129, 0.8265, 1.8577, 1.0867],
[0.8722, 0.8766, 1.3802, 1.0910], [0.7127, 0.5888, 1.6141, 1.0923]])
```

```
mat1 * mat1
tensor([[0.8103, 0.8442, 0.0192, 0.1370], [0.0379, 0.0514, 0.9192, 0.0684],
[0.5265, 0.5888, 0.3211, 0.0347], [0.2155, 0.1613, 0.9963, 0.7143]])
```

Recipe 2-7. Distributions
Problem

Knowledge of statistical distributions is essential for weight normalization, weight initialization, and computation of gradients in neural network–based operations using PyTorch. How do you know which distributions to use and when to use them?

Solution

Each statistical distribution follows a preestablished mathematical formula. You are going to use the most commonly used statistical distributions and their arguments in scenarios of problems.

How It Works

Bernoulli distribution is a special case of *binomial distribution* in which the number of trials can be more than one, but in a Bernoulli distribution, the number remains one. It is a discrete probability distribution of a random variable, which takes a value of 1 when there is probability that an event is a success and takes a value of 0 when there is probability that an event is a failure. A perfect example of this is tossing a coin, where 1 is heads and 0 is tails. Let's look at the program.

```
# about Bernoulli distribution
from torch.distributions.bernoulli import Bernoulli
dist = Bernoulli(torch.tensor([0.3,0.6,0.9]))
dist.sample() #sample is binary, it takes 1 with p and 0 with 1-p
tensor([0., 1., 0.])

#Creates a Bernoulli distribution parameterized by probs
#Samples are binary (0 or 1). They take the value 1 with probability p
#and 0 with probability 1 - p.
```

The *beta distribution* is a family of continuous random variables defined in the range of 0 and 1. This distribution is typically used for Bayesian inference analysis.

```
from torch.distributions.beta import Beta
dist = Beta(torch.tensor([0.5]), torch.tensor([0.5]))
dist
dist.sample()
```

The binomial distribution is applicable when the outcome is twofold and the experiment is repetitive. It belongs to the family of discrete probability distribution, where the probability of success is defined as 1 and the probability of failure is 0. The binomial distribution is used to model the number of successful events over many trials.

```
from torch.distributions.binomial import Binomial
dist = Binomial(100, torch.tensor([0 , .2, .8, 1]))
dist.sample()
tensor([ 0., 21., 83., 100.])

# 100- count of trials
# 0, 0.2, 0.8 and 1 are event probabilities
```

In probability and statistics, a categorical distribution can be defined as a generalized Bernoulli distribution, which is a discrete probability distribution that explains the possible results of any random variable that may take on one of the possible categories, with the probability of each category exclusively specified in the tensor.

```
from torch.distributions.categorical import Categorical
dist = Categorical(torch.tensor([ 0.20, 0.20, 0.20, 0.20, 0.20 ]))
dist
Categorical(probs: torch.Size([5]))
dist.sample()
tensor(2)
these are 0.20, 0.20, 0.20, 0.20,0.20 event probabilities.
```

A *Laplacian distribution* is a continuous probability distribution function that is otherwise known as a *double exponential distribution*. A Laplacian distribution is used in speech recognition systems to understand prior probabilities. It is also useful in Bayesian regression for deciding prior probabilities.

```
Laplace distribution parameterized by loc and 'scale'. Loc parameter is
mean or location parameter and scale is standard deviation parameter.
```

```
from torch.distributions.laplace import Laplace
dist = Laplace(torch.tensor([10.0]), torch.tensor([0.990]))
dist
Laplace(loc: tensor([10.]), scale: tensor([0.9900]))

dist.sample()
tensor([9.6554])
```

A *normal distribution* is very useful because of the property of the central limit theorem. It is defined by mean and standard deviations. If you know the mean and standard deviation of the distribution, you can estimate the event probabilities. See Figure 2-2.

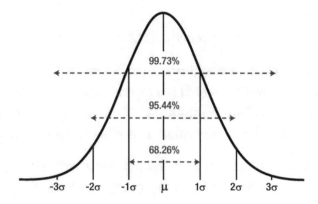

Figure 2-2. *Normal probability distribution*

```
#Normal (Gaussian) distribution parameterized by loc and 'scale'.
from torch.distributions.normal import Normal
dist = Normal(torch.tensor([100.0]), torch.tensor([10.0]))
dist
Normal(loc: tensor([100.]), scale: tensor([10.]))
dist.sample()
tensor([84.3435])
```

Conclusion

This chapter discussed sampling distribution and generating random numbers from distributions. Neural networks are the primary focus in tensor-based operations. Any sort of machine learning or deep learning model implementation requires gradient computation, updating weight, computing bias, and continuously updating the bias.

This chapter also discussed the statistical distributions supported by PyTorch and the situations where each type of distribution can be applied.

The next chapter discusses deep learning models in detail. These deep learning models include convolutional neural networks, recurrent neural networks, deep neural networks, and autoencoder models.

CHAPTER 3

CNN and RNN Using PyTorch

This chapter covers convolutional neural networks (CNN) and recurrent neural network and their implementation using PyTorch. Convolutional neural network is being used in image classification, object detection, object classification related tasks. The large scale image classification models requires PyTorch framework as it is considered to be faster than other frameworks, similarly the recurrent neural networks are used in natural language processing tasks such as text classification, sentiment classification, topic classification, audio classification etc. This chapter shows a few recipes on setting up CNN and RNN models, selecting the optimization function, saving a model, loading a model already trained and using the model for inference generation etc.

Recipe 3-1. Setting Up a Loss Function
Problem

How do you set up a loss function and optimize it? Choosing the right loss function increases the chances of model convergence.

Solution

In this recipe, you use another tensor as the update variable and introduce the tensors to the sample model and compute the error or loss. Then you compute the rate of change in the loss function to measure the choice of loss function in model convergence.

© Pradeepta Mishra 2023
P. Mishra, *PyTorch Recipes*, https://doi.org/10.1007/978-1-4842-8925-9_3

How It Works

In the following example, t_c and t_u are two tensors. This can be constructed from any NumPy array.

```
import torch
torch.__version__
'1.12.1+cu113'
torch.tensor
```

```
t_c = torch.tensor([0.5, 14.0, 15.0, 28.0, 11.0, 8.0, 3.0, -4.0, 6.0,
13.0, 21.0])
t_u = torch.tensor([35.7, 55.9, 58.2, 81.9, 56.3, 48.9, 33.9, 21.8, 48.4,
60.4, 68.4])
```

The sample model is just a linear equation to make the calculation happen. and the loss function defined if the mean square error (MSE) is shown next. Going forward in this chapter, you will increase the complexity of the model. For now, this is just a simple linear equation computation.

```
#height of people
t_c = torch.tensor([58.0, 59.0, 60.0, 61.0, 62.0, 63.0, 64.0, 65.0, 66.0,
67.0, 68.0, 69.0, 70.0, 71.0, 72.0])
```

```
#weight of people
t_u = torch.tensor([115.0, 117.0, 120.0, 123.0, 126.0, 129.0, 132.0, 135.0,
139.0, 142.0, 146.0, 150.0, 154.0, 159.0,164.0])
```

Let's now define the model. The w parameter is the weight tensor, which is multiplied with the t_u tensor. The result is added with a constant tensor, b, and the loss function chosen is a custom-built one; it is also available in PyTorch. In the following example, t_u is the tensor used, t_p is the tensor predicted, and t_c is the precomputed tensor, with which the predicted tensor needs to be compared to calculate the loss function.

```
def model(t_u, w, b):
    return w * t_u + b
```

```
def loss_fn(t_p, t_c):
    squared_diffs = (t_p - t_c)**2
    return squared_diffs.mean()
```

The formula w * t_u + b is the linear equation representation of a tensor-based computation.

```
w = torch.ones(1)
b = torch.zeros(1)

t_p = model(t_u, w, b)
t_p
tensor([115., 117., 120., 123., 126., 129., 132., 135., 139., 142., 146.,
150., 154., 159., 164.])

loss = loss_fn(t_p, t_c)
loss
tensor(5259.7334)
```

The initial loss value is 5259.7334, which is too high because of the initial round of weights chosen. The error in the first round of iteration is backpropagated to reduce the errors in the second round, for which the initial set of weights needs to be updated. Therefore, the rate of change in the loss function is essential in updating the weights in the estimation process.

```
delta = 0.1

loss_rate_of_change_w = (loss_fn(model(t_u,
                                        w + delta, b),
                            t_c) - loss_fn(model(t_u, w - delta, b),
                                        t_c)) / (2.0 * delta)
learning_rate = 1e-2

w = w - learning_rate * loss_rate_of_change_w
loss_rate_of_change_b = (loss_fn(model(t_u, w, b + delta), t_c) -
                        loss_fn(model(t_u, w, b - delta), t_c)) /
                        (2.0 * delta)

b = b - learning_rate * loss_rate_of_change_b
b
tensor([544.])
```

There are two parameters to update the rate of loss function: the learning rate at the current iteration and the learning rate at the previous iteration. If the delta between the two iterations exceeds a certain threshold, then the weight tensor needs to be updated or model convergence could happen. The preceding script shows the delta and learning rate values. Currently, they are static values that the user has the option to change.

This is how a simple mean square loss function works in a two-dimensional tensor example, with a tensor size of 10, 5.

Let's look at the following example. The MSELoss function is within the neural network module of PyTorch.

```
from torch import nn
loss = nn.MSELoss()
input = torch.randn(10, 5, requires_grad=True)
target = torch.randn(10, 5)
output = loss(input, target)
output.backward()
```

When you look at the gradient calculation that is used for backpropagation, it is shown as MSELoss.

```
output.grad_fn
<MseLossBackward0 at 0x7f424abd0f50>
```

Recipe 3-2. Estimating the Derivative of the Loss Function

Problem

How do you estimate the derivative of a loss function?

Solution

Using the following example, you change the loss function to two times the differences between the input and the output tensors, instead of the MSELoss function. The following grad_fn, which is defined as a custom function, shows how the final output retrieves the derivative of the loss function.

How It Works

Let's look at the following example. In the previous recipe, the last line of the script shows the grad_fn as an object embedded in the output object tensor. This recipe explains how it is computed. grad_fn is a derivative of the loss function with respect to the parameters of the model. This is exactly what you do in the following grad_fn:

```
input
tensor([[-1.0665, -0.8880, -1.1156, -0.1595, 0.2342], [ 3.1369, -0.6062,
1.0556, 1.9240, 1.0309], [ 1.8270, -0.8902, 1.8918, 1.0523, 1.8231],
[ 0.0969, 0.0462, -1.6298, -1.6399, 0.0167], [ 0.6968, -0.1793, 0.5698,
0.8613, -1.8561], [ 0.7462, -0.1504, 0.0779, 2.0298, 1.2302], [-2.1399,
-2.0118, -0.5827, 0.1486, 2.2127], [-0.2679, -0.5797, 0.5805, -0.4121,
0.5089], [ 1.3931, -0.8098, -0.3136, 0.6375, 0.6038], [ 1.8502, 0.0844,
0.7034, -0.1410, 2.5020]], requires_grad=True)
```

```
target
tensor([[ 2.1635, 2.8280, -1.2495, 0.3782, 0.7208], [ 0.4892, 0.4965,
-0.1423, 0.4918, -1.0321], [-1.4843, 0.4281, 0.6281, -1.4526, -1.8356],
[ 0.7769, 2.5248, -0.4420, 0.4313, -1.0156], [-1.4197, -1.0438, 1.0570,
0.3100, 0.6264], [-1.3284, -0.9601, 0.0358, 0.5170, 1.5762], [ 0.2165,
-1.0205, 0.2125, 0.4595, 0.9997], [ 0.4572, -0.3321, 0.3248, -0.4419,
-0.0550], [-1.6006, 0.4164, -0.5147, -1.0651, -1.8708], [-1.9251, 0.9669,
-0.9007, -0.4605, -0.1377]])
```

```python
def dloss_fn(t_p, t_c):
    dsq_diffs = 2 * (t_p - t_c)
    return dsq_diffs

def model(t_u, w, b):
    return w * t_u + b

def dmodel_dw(t_u, w, b):
    return t_u

def dmodel_db(t_u, w, b):
    return 1.0
```

```
def grad_fn(t_u, t_c, t_p, w, b):
    dloss_dw = dloss_fn(t_p, t_c) * dmodel_dw(t_u, w, b)
    dloss_db = dloss_fn(t_p, t_c) * dmodel_db(t_u, w, b)
    return torch.stack([dloss_dw.mean(), dloss_db.mean()])
```

The parameters are the input, bias settings, learning rate, and the number of epochs for the model training. The estimation of these parameters provides values to the equation.

```
params = torch.tensor([1.0, 0.0])

nepochs = 10

learning_rate = 0.005

for epoch in range(nepochs):
    # forward pass
    w, b = params
    t_p = model(t_u, w, b)

    loss = loss_fn(t_p, t_c)
    print('Epoch %d, Loss %f' % (epoch, float(loss)))

    # backward pass
    grad = grad_fn(t_u, t_c, t_p, w, b)

    print('Params:', params)
    print('Grad:', grad)

    params = params - learning_rate * grad

params
Epoch 0, Loss 5259.733398
Params: tensor([1., 0.])
Grad: tensor([19936.2676,    143.4667])
Epoch 1, Loss 186035504.000000
Params: tensor([-98.6813,   -0.7173])
Grad: tensor([-3752242.2500,    -27117.4902])
Epoch 2, Loss 6590070521856.000000
Params: tensor([18662.5293,    134.8701])
Grad: tensor([7.0622e+08, 5.1037e+06])...............
```

This is what the initial result looks like. Epoch is an iteration that produces a loss value from the loss function defined earlier. The parameters vector is about coefficients and constants that need to be changed to minimize the loss function. The grad function computes the feedback value to the next epoch. This is just an example. The number of epochs chosen is an iterative task depending on the input data, output data, and choice of loss and optimization functions.

If you reduce the learning rate, you are able to pass relevant values to the gradient, the parameter updates in a better way, and model convergence is achieved within few iterations.

```
params = torch.tensor([1.0, 0.0])

nepochs = 10

learning_rate = 0.1

for epoch in range(nepochs):
    # forward pass
    w, b = params
    t_p = model(t_u, w, b)

    loss = loss_fn(t_p, t_c)
    print('Epoch %d, Loss %f' % (epoch, float(loss)))

    # backward pass
    grad = grad_fn(t_u, t_c, t_p, w, b)

    print('Params:', params)
    print('Grad:', grad)

    params = params - learning_rate * grad

params
```

```
Epoch 0, Loss 5259.733398
Params: tensor([1., 0.])
Grad: tensor([19936.2676,    143.4667])
Epoch 1, Loss 75167318016.000000
Params: tensor([-1992.6268,    -14.3467])
Grad: tensor([-75423624.0000,    -545075.6875])
```

```
Epoch 2, Loss 1075861270101491712.000000
Params: tensor([7540370.0000,   54493.2227])
Grad: tensor([2.8535e+11, 2.0621e+09])................
```

If you reduce the learning rate a bit, the process of weight updating will be a little slower, which means that the epoch number needs to be increased in order to find a stable state for the model.

```
t_un = 0.1 * t_u
params = torch.tensor([1.0, 0.0])

nepochs = 10

learning_rate = 0.05

for epoch in range(nepochs):
    # forward pass
    w, b = params
    t_p = model(t_un, w, b)

    loss = loss_fn(t_p, t_c)
    print('Epoch %d, Loss %f' % (epoch, float(loss)))

    # backward pass
    grad = grad_fn(t_un, t_c, t_p, w, b)

    print('Params:', params)
    print('Grad:', grad)

    params = params - learning_rate * grad

params
```

The following are the results:

```
Epoch 0, Loss 2642.455322
Params: tensor([1., 0.])
Grad: tensor([-1412.0094,  -102.6533])
Epoch 1, Loss 855426.562500
Params: tensor([71.6005,  5.1327])
Grad: tensor([25443.8555,  1838.2997])
```

```
Epoch 2, Loss 277741792.000000
Params: tensor([-1200.5923,    -86.7823])
Grad: tensor([-458472.5938,  -33135.7656]).................
```

If you increase the number of epochs, then what happens to the loss function and parameter tensor can be viewed in the following script, in which you print the loss value to find the minimum loss corresponding to the epoch. Then you can extract the best parameters from the model.

```
params = torch.tensor([1.0, 0.0])

nepochs = 50

learning_rate = 1e-2

for epoch in range(nepochs):
    # forward pass
    w, b = params
    t_p = model(t_un, w, b)

    loss = loss_fn(t_p, t_c)
    print('Epoch %d, Loss %f' % (epoch, float(loss)))

    # backward pass
    grad = grad_fn(t_un, t_c, t_p, w, b)

    params = params - learning_rate * grad

params
```

The following are the results:

```
Epoch 0, Loss 2642.455322
Epoch 1, Loss 20719.347656
Epoch 2, Loss 162827.593750
Epoch 3, Loss 1279985.125000
Epoch 4, Loss 10062318.000000
Epoch 5, Loss 79103048.000000
Epoch 6, Loss 621853952.000000
Epoch 7, Loss 4888591872.000000
Epoch 8, Loss 38430781440.000000
```

The following is the final loss value at the final epoch level. This is called an exploding gradient problem and it happens due to bad initialization or an incorrect learning rate or both. To address this, either you need to initialize with clipping or apply clipping on the gradients calculation.

```
Epoch 37, Loss 358133256611123759997199585901753139 2.000000
Epoch 38, Loss inf
Epoch 39, Loss inf
Epoch 40, Loss inf
Epoch 41, Loss inf
Epoch 42, Loss inf
Epoch 43, Loss inf
Epoch 44, Loss inf
Epoch 45, Loss inf
Epoch 46, Loss inf
Epoch 47, Loss inf
Epoch 48, Loss inf
Epoch 49, Loss inf
tensor([-9.0577e+22, -6.5463e+21])
```

To fine-tune this model in estimating parameters, you can redefine the model and the loss function and apply it to the same example.

```
def model(t_u, w, b):
    return w * t_u + b

def loss_fn(t_p, t_c):
    sq_diffs = (t_p - t_c)**2
    return sq_diffs.mean()
```

Set up the parameters. After completing the training process, you should reset the grad function to None.

```
params = torch.tensor([1.0, 0.0], requires_grad=True)

loss = loss_fn(model(t_u, *params), t_c)
params.grad is None
```

Recipe 3-3. Fine-Tuning a Model

Problem

How do you find the gradients of the loss function by applying an optimization function to optimize the loss function?

Solution

You'll use the backward() function.

How It Works

Let's look at the following example. The backward() function calculates the gradients of a function with respect to its parameters. In this section, you retrain the model with new set of hyperparameters.

```
loss.backward()
params.grad
tensor([19936.2676, 143.4667])
```

Reset the parameter grid. If you do not reset the parameters in an existing session, the error values accumulated from any other session become mixed, so it is important to reset the parameter grid.

```
if params.grad is not None:
    params.grad.zero_()

def model(t_u, w, b):
    return w * t_u + b

def loss_fn(t_p, t_c):
    sq_diffs = (t_p - t_c)**2
    return sq_diffs.mean()
```

After redefining the model and the loss function, let's retrain the model.

```
params = torch.tensor([1.0, 0.0], requires_grad=True)
```

```
nepochs = 5000

learning_rate = 1e-2
for epoch in range(nepochs):
    # forward pass
    t_p = model(t_un, *params)
    loss = loss_fn(t_p, t_c)

    print('Epoch %d, Loss %f' % (epoch, float(loss)))

    # backward pass
    if params.grad is not None:
        params.grad.zero_()

    loss.backward()

    #params.grad.clamp_(-1.0, 1.0)
    #print(params, params.grad)

    params = (params - learning_rate * params.grad).detach().
    requires_grad_()

params
```

Recipe 3-4. Selecting an Optimization Function

Problem

How do you optimize the gradients with the function in Recipe 3-3?

Solution

There are certain functions that are embedded in PyTorch, and there are certain
optimization functions that the user has to create.

How It Works

Let's look at the following example:

```
import torch.optim as optim

dir(optim)
['ASGD', 'Adadelta', 'Adagrad', 'Adam', 'AdamW', 'Adamax', 'LBFGS',
'NAdam', 'Optimizer', 'RAdam', 'RMSprop', 'Rprop', 'SGD', 'SparseAdam',
'__builtins__', '__cached__', '__doc__', '__file__', '__loader__',
'__name__', '__package__', '__path__', '__spec__', '_functional',
'_multi_tensor', 'lr_scheduler', 'swa_utils']
```

Each optimization method is unique in solving a problem. I will describe this later.

The Adam optimizer is a first-order, gradient-based optimization of stochastic objective functions. It is based on adaptive estimation of lower-order moments. This is computationally efficient enough for deployment on large datasets. To use `torch.optim`, you must construct an optimizer object in your code that will hold the current state of the parameters and will update the parameters based on the computed gradients, moments, and learning rate. To construct an optimizer, you must give it an iterable containing the parameters and ensure that all the parameters are variables to optimize. Then, you can specify optimizer-specific options, such as the learning rate, weight decay, moments, and so forth.

SGD is another optimizer that is fast enough to work on large datasets. This method does not require manual fine-tuning of the learning rate; the algorithm takes care of it internally.

```
params = torch.tensor([1.0, 0.0], requires_grad=True)

learning_rate = 1e-5

optimizer = optim.SGD([params], lr=learning_rate)
t_p = model(t_u, *params)

loss = loss_fn(t_p, t_c)

loss.backward()

optimizer.step()
```

```
params
tensor([ 0.8006, -0.0014], requires_grad=True)

params = torch.tensor([1.0, 0.0], requires_grad=True)

learning_rate = 1e-2

optimizer = optim.SGD([params], lr=learning_rate)

t_p = model(t_un, *params)

loss = loss_fn(t_p, t_c)

optimizer.zero_grad()

loss.backward()

optimizer.step()

params
tensor([15.1201, 1.0265], requires_grad=True)
```

Now let's call the model and loss function again and apply them along with the optimization function.

```
def model(t_u, w, b):
    return w * t_u + b

def loss_fn(t_p, t_c):
    sq_diffs = (t_p - t_c)**2
    return sq_diffs.mean()

params = torch.tensor([1.0, 0.0], requires_grad=True)

nepochs = 5000
learning_rate = 1e-2

optimizer = optim.SGD([params], lr=learning_rate)

for epoch in range(nepochs):

    # forward pass
    t_p = model(t_un, *params)
    loss = loss_fn(t_p, t_c)
```

```
    print('Epoch %d, Loss %f' % (epoch, float(loss)))

    # backward pass
    optimizer.zero_grad()
    loss.backward()
    optimizer.step()

t_p = model(t_un, *params)

params
```

Let's look at the gradient in a loss function. Using the optimization library, you can try to find the best value of the loss function.

The example has two custom functions and a loss function. You have taken two small tensor values. The new thing is that you have used the optimizer to find the minimum value.

In the following example, you use Adam as the optimizer:

```
def model(t_u, w, b):
    return w * t_u + b

def loss_fn(t_p, t_c):
    sq_diffs = (t_p - t_c)**2
    return sq_diffs.mean()

params = torch.tensor([1.0, 0.0], requires_grad=True)

nepochs = 5000
learning_rate = 1e-1

optimizer = optim.Adam([params], lr=learning_rate)

for epoch in range(nepochs):
    # forward pass
    t_p = model(t_u, *params)
    loss = loss_fn(t_p, t_c)

    print('Epoch %d, Loss %f' % (epoch, float(loss)))
```

```
    # backward pass
    optimizer.zero_grad()
    loss.backward()
    optimizer.step()

t_p = model(t_u, *params)

params

Epoch 0, Loss 5259.733398
Epoch 1, Loss 3443.706543
Epoch 2, Loss 2025.263306
Epoch 3, Loss 1002.202881
Epoch 4, Loss 357.638672
Epoch 5, Loss 53.362339
Epoch 6, Loss 24.627544
Epoch 7, Loss 181.475266
Epoch 8, Loss 421.412598
Epoch 9, Loss 651.219666
Epoch 10, Loss 806.726135
```

In the preceding code, you computed the optimized parameters and computed the predicted tensors using the actual and predicted tensors. You can display a graph that has a line shown as a regression line.

```
Epoch 4993, Loss 0.167906
Epoch 4994, Loss 0.167906
Epoch 4995, Loss 0.167905
Epoch 4996, Loss 0.167904
Epoch 4997, Loss 0.167903
Epoch 4998, Loss 0.167903
Epoch 4999, Loss 0.167903
tensor([ 0.2879, 25.6386], requires_grad=True)
```

Let's visualize the sample data in graphical form using the actual and predicted tensors. See Figure 3-1.

```
from matplotlib import pyplot as plt
%matplotlib inline
```

```
plt.plot(0.1 * t_u.numpy(), t_p.detach().numpy())
plt.plot(0.1 * t_u.numpy(), t_c.numpy(), 'o')
```

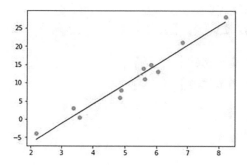

Figure 3-1. *Actual vs. Predicted Tensor*

Recipe 3-5. Further Optimizing the Function

Problem

How do you optimize the training set and test it with a validation set using random samples?

Solution

Here's the process of further optimization.

How It Works

Let's look at the following example. Here you set the number of samples and then you take 20% of the data as validation samples using shuffled_indices. You take random samples of all the records. The objective of the train and validation set is to build a model in a training set, make the prediction on the validation set, and check the accuracy of the model.

```
n_samples = t_u.shape[0]
n_val = int(0.2 * n_samples)

shuffled_indices = torch.randperm(n_samples)
```

```
train_indices = shuffled_indices[:-n_val]
val_indices = shuffled_indices[-n_val:]

train_indices, val_indices
(tensor([14, 13, 7, 6, 11, 0, 10, 3, 4, 5, 12, 1]), tensor([8, 2, 9]))

t_u_train = t_u[train_indices]
t_c_train = t_c[train_indices]

t_u_val = t_u[val_indices]
t_c_val = t_c[val_indices]

def model(t_u, w, b):
    return w * t_u + b

def loss_fn(t_p, t_c):
    sq_diffs = (t_p - t_c)**2
    return sq_diffs.mean()

params = torch.tensor([1.0, 0.0], requires_grad=True)

nepochs = 5000
learning_rate = 1e-2

optimizer = optim.SGD([params], lr=learning_rate)

t_un_train = 0.1 * t_u_train
t_un_val = 0.1 * t_u_val
```

Now let's run the train and validation process. You first take the training input data and multiply it by the parameter's next line. You make a prediction and compute the loss function. Using the same model in the third line, you make predictions and then you evaluate the loss function for the validation dataset. In the backpropagation process, you calculate the gradient of the loss function for the training set, and using the optimizer, you update the parameters.

```
for epoch in range(nepochs):

    # forward pass
    t_p_train = model(t_un_train, *params)
    loss_train = loss_fn(t_p_train, t_c_train)
```

```
    t_p_val = model(t_un_val, *params)
    loss_val = loss_fn(t_p_val, t_c_val)

    print('Epoch %d, Training loss %f, Validation loss %f' % (epoch,
    float(loss_train),

                                                              float
                                                              (loss_val)))

    # backward pass
    optimizer.zero_grad()
    loss_train.backward()
    optimizer.step()

t_p = model(t_un, *params)

params

Epoch 0, Training loss 2652.548340, Validation loss 2602.083252
Epoch 1, Training loss 21507.599609, Validation loss 19826.755859
Epoch 2, Training loss 174840.953125, Validation loss 164931.500000
Epoch 3, Training loss 1421780.000000, Validation loss 1330612.750000
Epoch 4, Training loss 11562152.000000, Validation loss 10851068.000000
Epoch 5, Training loss 94025800.000000, Validation loss 88156856.000000
Epoch 6, Training loss 764637760.000000, Validation loss 717156288.000000
Epoch 7, Training loss 6218196992.000000, Validation loss 5831366144.000000
Epoch 8, Training loss 50567704576.000000, Validation loss
47423901696.000000
```

The following are the last 10 epochs and their results:

```
Epoch 4989, Training loss nan, Validation loss nan
Epoch 4990, Training loss nan, Validation loss nan
Epoch 4991, Training loss nan, Validation loss nan
Epoch 4992, Training loss nan, Validation loss nan
Epoch 4993, Training loss nan, Validation loss nan
Epoch 4994, Training loss nan, Validation loss nan
Epoch 4995, Training loss nan, Validation loss nan
Epoch 4996, Training loss nan, Validation loss nan
Epoch 4997, Training loss nan, Validation loss nan
```

```
Epoch 4998, Training loss nan, Validation loss nan
Epoch 4999, Training loss nan, Validation loss nan
tensor([nan, nan], requires_grad=True)
```

In the previous step, the gradient was set to true. In the following set, you disable gradient calculation by using the torch.no_grad() function. The rest of the syntax remains the same. Disabling the gradient calculation is useful for drawing inferences when you are sure that you will not call Tensor.backward(). This reduces memory consumption for computations that would otherwise need requires_grad=True.

```
for epoch in range(nepochs):

    # forward pass
    t_p_train = model(t_un_train, *params)
    loss_train = loss_fn(t_p_train, t_c_train)

    with torch.no_grad():
        t_p_val = model(t_un_val, *params)
        loss_val = loss_fn(t_p_val, t_c_val)

    print('Epoch %d, Training loss %f, Validation loss %f' % (epoch,
    float(loss_train),

                                                        float
                                                        (loss_val)))

    # backward pass
    optimizer.zero_grad()
    loss_train.backward()
    optimizer.step()

params
Epoch 0, Training loss nan, Validation loss nan
Epoch 1, Training loss nan, Validation loss nan
Epoch 2, Training loss nan, Validation loss nan
Epoch 3, Training loss nan, Validation loss nan
Epoch 4, Training loss nan, Validation loss nan
Epoch 5, Training loss nan, Validation loss nan
Epoch 6, Training loss nan, Validation loss nan
Epoch 7, Training loss nan, Validation loss nan
```

```
Epoch 8, Training loss nan, Validation loss nan
Epoch 9, Training loss nan, Validation loss nan
Epoch 10, Training loss nan, Validation loss nan............
```

The last rounds of epochs are displayed in other lines of code, as follows:

The final parameters are 5.44 and –18.012.

Recipe 3-6. Implementing a Convolutional Neural Network (CNN)

Problem

How do you implement a convolutional neural network using PyTorch?

Solution

There are various built-in datasets available on torchvision. You are considering the MNIST dataset and trying to build a CNN model.

How It Works

Let's look at the following example. As a first step, you set up the hyperparameters. The second step is to set up the architecture. The last step is to train the model and make predictions.

```
import torch
import torch.nn as nn
from torch.autograd import Variable
import torch.utils.data as Data
import torchvision
import matplotlib.pyplot as plt
%matplotlib inline
torch.manual_seed(1)     # reproducible
```

In the preceding code, you import the necessary libraries for deploying the convolutional neural network model using the digits dataset (Figure 3-2). The MNIST digits dataset is the most popular dataset in deep learning for computer vision and image processing.

```
# Hyper Parameters
EPOCH = 1
# train the input data n times, to save time, we just train 1 epoch
BATCH_SIZE = 50
# 50 samples at a time to pass through the epoch
LR = 0.001
# learning rate
DOWNLOAD_MNIST = True
# set to False if you have downloaded
# Mnist digits dataset
train_data = torchvision.datasets.MNIST(
    root='./mnist/',
    train=True,
    # this is training data
    transform=torchvision.transforms.ToTensor(),
    # torch.FloatTensor of shape (Color x Height x Width) and
    #normalize in the range [0.0, 1.0]
    download=DOWNLOAD_MNIST,
    # download it if you don't have it
)

# plot one example
print(train_data.train_data.size())              # (60000, 28, 28)
print(train_data.train_labels.size())            # (60000)
plt.imshow(train_data.train_data[0].numpy(), cmap='gray')
plt.title('%i' % train_data.train_labels[0])
plt.show()
torch.Size([60000, 28, 28])
torch.Size([60000])
```

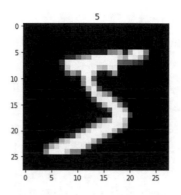

Figure 3-2.

Let's load the dataset using the loader functionality

```
# Data Loader for easy mini-batch return in training, the image batch
shape will be
#(50, 1, 28, 28)
train_loader = Data.DataLoader(dataset=train_data, batch_size=BATCH_SIZE,
shuffle=True)
```

```
# convert test data into Variable, pick 2000 samples to speed up testing
test_data = torchvision.datasets.MNIST(root='./mnist/', train=False)
test_x = Variable(torch.unsqueeze(test_data.test_data, dim=1)).type(torch.
FloatTensor)[:2000]/255.
# shape from (2000, 28, 28) to (2000, 1, 28, 28), value in range(0,1)
test_y = test_data.test_labels[:2000]
```

In convolutional neural network architecture, the input image is converted to
a feature set as set by color times height and width of the image. Because of the
dimensionality of the dataset, you cannot model it to predict the output. The output
layer in the preceding graph has classes such as car, truck, van, and bicycle. The input
bicycle image has features that the CNN model should make use of and predict correctly.
The convolution layer is usually accompanied by the pooling layer, which can be max
pooling and average pooling. The different layers of pooling and convolution continue
until the dimensionality is reduced to a level where you can use fully connected simple
neural networks to predict the correct classes. See Figure 3-3.

```
class CNN(nn.Module):
    def __init__(self):
```

```
        super(CNN, self).__init__()
        self.conv1 = nn.Sequential(          # input shape (1, 28, 28)
            nn.Conv2d(
                in_channels=1,               # input height
                out_channels=16,             # n_filters
                kernel_size=5,               # filter size
                stride=1,                    # filter movement/step
                padding=2,
                # if want same width and length of this image after con2d,
                #padding=(kernel_size-1)/2 if stride=1
            ),                               # output shape (16, 28, 28)
            nn.ReLU(),                       # activation
            nn.MaxPool2d(kernel_size=2),
            # choose max value in 2x2 area, output shape (16, 14, 14)
        )
        self.conv2 = nn.Sequential(          # input shape (1, 28, 28)
            nn.Conv2d(16, 32, 5, 1, 2),      # output shape (32, 14, 14)
            nn.ReLU(),                       # activation
            nn.MaxPool2d(2),                 # output shape (32, 7, 7)
        )
        self.out = nn.Linear(32 * 7 * 7, 10)
        # fully connected layer, output 10 classes

    def forward(self, x):
        x = self.conv1(x)
        x = self.conv2(x)
        x = x.view(x.size(0), -1)
        # flatten the output of conv2 to (batch_size, 32 * 7 * 7)
        output = self.out(x)
        return output, x     # return x for visualization

cnn = CNN()
print(cnn)  # net architecture

CNN(
  (conv1): Sequential(
    (0): Conv2d(1, 16, kernel_size=(5, 5), stride=(1, 1), padding=(2, 2))
```

```
    (1): ReLU()
    (2): MaxPool2d(kernel_size=2, stride=2, padding=0, dilation=1,
        ceil_mode=False)
  )
  (conv2): Sequential(
    (0): Conv2d(16, 32, kernel_size=(5, 5), stride=(1, 1), padding=(2, 2))
    (1): ReLU()
    (2): MaxPool2d(kernel_size=2, stride=2, padding=0, dilation=1,
        ceil_mode=False)
  )
  (out): Linear(in_features=1568, out_features=10, bias=True)
)
optimizer = torch.optim.Adam(cnn.parameters(), lr=LR)   # optimize all cnn
                                                          parameters
loss_func = nn.CrossEntropyLoss()                        # the target label
                                                          is not one-hotted

import sklearn
import warnings
warnings.filterwarnings("ignore", category=FutureWarning)
import warnings
warnings.filterwarnings("ignore")

from matplotlib import cm
try: from sklearn.manifold import TSNE; HAS_SK = True
except: HAS_SK = False; print('Please install sklearn for layer
visualization, if not there')
def plot_with_labels(lowDWeights, labels):
    plt.cla()
    X, Y = lowDWeights[:, 0], lowDWeights[:, 1]
    for x, y, s in zip(X, Y, labels):
        c = cm.rainbow(int(255 * s / 9)); plt.text(x, y, s,
        backgroundcolor=c, fontsize=9)
    plt.xlim(X.min(), X.max()); plt.ylim(Y.min(), Y.max()); plt.
    title('Visualize last layer');
    plt.show();
    #plt.pause(0.01)
```

```
plt.ion()
# training and testing
for epoch in range(EPOCH):
    for step, (x, y) in enumerate(train_loader):
        # gives batch data, normalize x when iterate train_loader
        b_x = Variable(x)    # batch x
        b_y = Variable(y)    # batch y

        output = cnn(b_x)[0]                 # cnn output
        loss = loss_func(output, b_y)   # cross entropy loss
        optimizer.zero_grad()           # clear gradients for this
                                        training step
        loss.backward()                 # backpropagation, compute
                                        gradients
        optimizer.step()                # apply gradients

        if step % 100 == 0:
            test_output, last_layer = cnn(test_x)
            pred_y = torch.max(test_output, 1)[1].data.squeeze()
            accuracy = (pred_y == test_y).sum().item() / float(test_y.
            size(0))
            print('Epoch: ', epoch, '| train loss: %.4f' % loss.data,
                  '| test accuracy: %.2f' % accuracy)
            if HAS_SK:
                # Visualization of trained flatten layer (T-SNE)
                tsne = TSNE(perplexity=30, n_components=2, init='pca',
                n_iter=5000)
                plot_only = 500
                low_dim_embs = tsne.fit_transform(last_layer.data.numpy()
                [:plot_only, :])
                labels = test_y.numpy()[:plot_only]
                plot_with_labels(low_dim_embs, labels)
plt.ioff()
```

In the preceding graph, if you look at the number 4, it is scattered throughout the graph. Ideally, all of the 4s are closer to each other. This is because the test accuracy was very low.

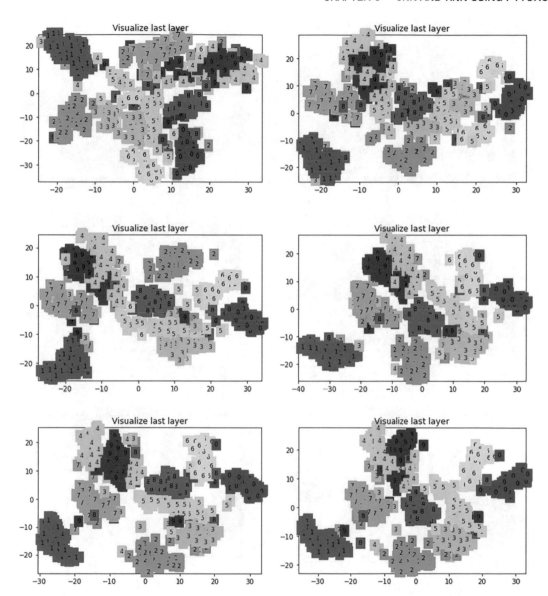

Figure 3-3.

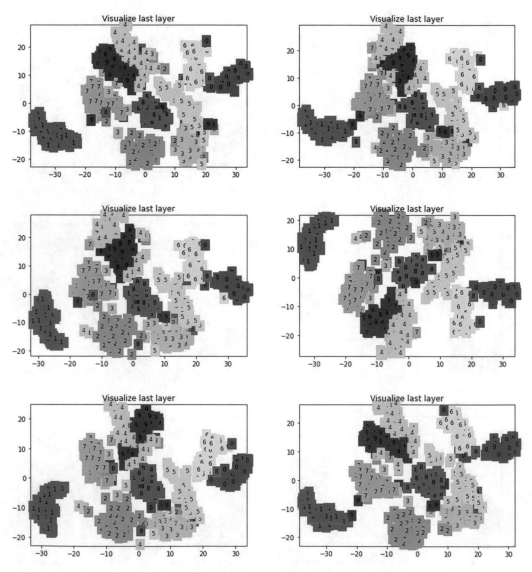

Figure 3-3. *(continued)*

In this iteration, the training loss is reduced from 0.4369 to 0.1482 and the test accuracy improves from 16% to 94%. The digits with the same color are placed closely on the graph.

In the next epoch, the test accuracy on the MNIST digits dataset the accuracy increases to 95%.

In the final step/epoch, the digits with similar numbers are placed together. After training a model successfully, the next step is to make use of the model to predict. The following code explains the predictions process. The output object is numbered as 0, 1, 2, and so forth. The following shows the real and predicted numbers.

```
# print 10 predictions from test data
test_output, _ = cnn(test_x[:10])
pred_y = torch.max(test_output, 1)[1].data.numpy().squeeze()
print(pred_y, 'prediction number')
print(test_y[:10].numpy(), 'real number')
```

Recipe 3-7. Reloading a Model
Problem

How do you store and reload a model that has already been trained? Given the nature of deep learning models, which typically require a larger training time, the computational process creates a huge cost to the company. Can you retrain the model with new inputs and store the model?

Solution

In the production environment, you typically cannot train and predict at the same time because the training process takes a very long time. The prediction services cannot be applied until the training process using the epoch is completed. Disassociating the training process from the prediction process is required; therefore, you need to store the application's trained model and continue until the next phase of training is done.

How It Works

Let's look at the following example, where you create the save function, which uses the Torch neural network module to create the model and the restore_net() function to get back the neural network model that was trained earlier.

```
import torch
from torch.autograd import Variable
import matplotlib.pyplot as plt
%matplotlib inline
```

```
torch.manual_seed(1)    # reproducible

#sample data
x = torch.unsqueeze(torch.linspace(-1, 1, 100), dim=1)  # x data (tensor),
                                                  shape=(100, 1)
y = x.pow(2) + 0.2*torch.rand(x.size())  # noisy y data (tensor),
                                      shape=(100, 1)
x, y = Variable(x, requires_grad=False), Variable(y, requires_grad=False)
```

The preceding script contains a dependent Y variable and an independent X variable as sample data points to create a neural network model. The following save function stores the model. The net1 object is the trained neural network model, which can be stored using two different protocols: (1) save the entire neural network model with all the weights and biases, and (2) save the model using only the weights. If the trained model object is very heavy in terms of size, you should save only the parameters that are weights; if the trained object size is low, then the entire model can be stored.

```
def save():
    # save net1
    net1 = torch.nn.Sequential(
        torch.nn.Linear(1, 10),
        torch.nn.ReLU(),
        torch.nn.Linear(10, 1)
    )
    optimizer = torch.optim.SGD(net1.parameters(), lr=0.5)
    loss_func = torch.nn.MSELoss()

    for t in range(100):
        prediction = net1(x)
        loss = loss_func(prediction, y)
        optimizer.zero_grad()
        loss.backward()
        optimizer.step()

    # plot result
    plt.figure(1, figsize=(10, 3))
```

```
plt.subplot(131)
plt.title('Net1')
plt.scatter(x.data.numpy(), y.data.numpy())
plt.plot(x.data.numpy(), prediction.data.numpy(), 'r-', lw=5)

# 2 ways to save the net
torch.save(net1, 'net.pkl')  # save entire net
torch.save(net1.state_dict(), 'net_params.pkl')   # save only the
                                                    parameters
```

The prebuilt neural network model can be reloaded to the existing PyTorch session by using the load function. To test the net1 object and make predictions, you load the net1 object and store the model as net2. By using the net2 object, you can predict the outcome variable. The following script generates the graph as a dependent and an independent variable. prediction.data.numpy() in the last line of the code shows the predicted result.

```
def restore_net():
    # restore entire net1 to net2
    net2 = torch.load('net.pkl')
    prediction = net2(x)

    # plot result
    plt.subplot(132)
    plt.title('Net2')
    plt.scatter(x.data.numpy(), y.data.numpy())
    plt.plot(x.data.numpy(), prediction.data.numpy(), 'r-', lw=5)
```

Loading the pickle file format of the entire neural network is a relatively slow process; however, if you are only making predictions for a new dataset, you can only load the parameters of the model in a pickle format rather than the whole network.

```
def restore_params():
    # restore only the parameters in net1 to net3
    net3 = torch.nn.Sequential(
        torch.nn.Linear(1, 10),
        torch.nn.ReLU(),
        torch.nn.Linear(10, 1)
    )
```

```
    # copy net1's parameters into net3
    net3.load_state_dict(torch.load('net_params.pkl'))
    prediction = net3(x)

    # plot result
    plt.subplot(133)
    plt.title('Net3')
    plt.scatter(x.data.numpy(), y.data.numpy())
    plt.plot(x.data.numpy(), prediction.data.numpy(), 'r-', lw=5)
    plt.show()
# save net1
save()
# restore entire net (may slow)
restore_net()
# restore only the net parameters
restore_params()
```

Reuse the model. The `restore` function makes sure that the trained parameters can be reused by the model. To restore the model, you can use the `load_state_dict()` function to load the parameters of the model. The three models in Figure 3-4 are identical because `net2` and `net3` are copies of `net1`.

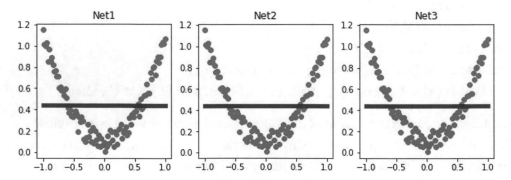

Figure 3-4.

Recipe 3-8. Implementing a Recurrent Neural Network

Problem

How do you set up a recurrent neural network (RNN) using the MNIST dataset?

Solution

A recurrent neural network is considered a memory network. You will use the epoch as 1 and a batch size of 64 samples at a time to establish the connection between the input and the output. Using the RNN model, you can predict the digits present in the images. See Figure 3-5.

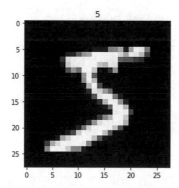

Figure 3-5.

How It Works

Let's look at the following example. The recurrent neural network takes a sequence of vectors in the input layer and produces a sequence of vectors in the output layer. The information sequence is processed through the internal state transfer in the recurrent layer. Sometimes the output values have a long dependency on past historical values. This is another variant of the RNN model: the long short-term memory (LSTM) model. It is applicable for any sort of domain where the information is consumed in a sequential manner, such as in a time series where the current stock price is decided by the historical stock price, where the dependency can be short or long. Similarly, the context prediction using the long and short range of textual input vectors. There are other industry use cases, such as noise classification, where noise is also a sequence of information.

The following code explains the execution of the RNN model using the PyTorch module. There are three sets of weights: U, V, and W. The set of weights vector, represented by W, is for passing information among the memory cells in the network that display communication among the hidden state. RNN uses an embedding layer using the Word2vec representation. The embedding matrix is the size of the number of words by the number of neurons in the hidden layer. If you have 20,000 words and 1,000 hidden units, for example, the matrix has a 20,000×1000 size of the embedding layer. The new representations are passed to the LSTM cells, which go to a sigmoid output layer.

```
import torch
from torch import nn
from torch.autograd import Variable
import torchvision.datasets as dsets
import torchvision.transforms as transforms
import matplotlib.pyplot as plt
%matplotlib inline
torch.manual_seed(1)      # reproducible

# Hyper Parameters
EPOCH = 1                 # train the training data n times, to save time, we
                          # just train 1 epoch
BATCH_SIZE = 64
TIME_STEP = 28            # rnn time step / image height
INPUT_SIZE = 28          # rnn input size / image width
LR = 0.01                # learning rate
DOWNLOAD_MNIST = True     # set to True if haven't download the data
```

The RNN models have hyperparameters, such as the number of iterations (EPOCH); the batch size dependent on the memory available in a single machine; a time step to remember the sequence of information; input size, which shows the vector size; and learning rate. The selection of these values is indicative; you cannot depend on them for other use cases. The value selection for hyperparameter tuning is an iterative process; either you can choose multiple parameters and decide which one is working or do parallel training of the model and decide which one is working fine.

```
# Mnist digital dataset
train_data = dsets.MNIST(
    root='./mnist/',
    train=True,                          # this is training data
    transform=transforms.ToTensor(),     # Converts a PIL.Image or numpy.
                                         ndarray to
                                         # torch.FloatTensor of shape
                                         (C x H x W) and normalize in
                                         the range [0.0, 1.0]
    download=DOWNLOAD_MNIST,             # download it if you don't have it
)

# plot one example
print(train_data.train_data.size())     # (60000, 28, 28)
print(train_data.train_labels.size())   # (60000)
plt.imshow(train_data.train_data[0].numpy(), cmap-'gray')
plt.title('%i' % train_data.train_labels[0])
plt.show()
torch.Size([60000, 28, 28])
torch.Size([60000])
```

Using the dsets.MINIST() function, you can load the dataset to the current session. If you need to store the dataset, then download it locally.

The preceding script shows what the sample image dataset looks like. To train the deep learning model, you need to convert the whole training dataset into mini batches, which help you with averaging the final accuracy of the model. By using the data loader function, you can load the training data and prepare the mini batches. The purpose of the shuffle selection in mini batches is to ensure that the model captures all the variations in the actual dataset.

```
# Data Loader for easy mini-batch return in training
train_loader = torch.utils.data.DataLoader(dataset=train_data,
                                           batch_size=BATCH_SIZE,
                                           shuffle=True)

# convert test data into Variable, pick 2000 samples to speed up testing
test_data = dsets.MNIST(root='./mnist/', train=False, transform=transforms.
ToTensor())
```

```
test_x = Variable(test_data.test_data, volatile=True).type(torch.
FloatTensor)[:2000]/255.
# shape (2000, 28, 28) value in range(0,1)
test_y = test_data.test_labels.numpy().squeeze()[:2000]    # covert to
                                                              numpy array
```

The preceding script prepares the training dataset. The test data is captured with the flag train=False. It is transformed to a tensor using the test data random sample of 2,000 each at a time is picked up for testing the model. The test features set is converted to a variable format and the test label vector is represented in a NumPy array format.

```
class RNN(nn.Module):
    def __init__(self):
        super(RNN, self).__init__()

        self.rnn = nn.LSTM(          # if use nn.RNN(), it hardly learns
            input_size=INPUT_SIZE,
            hidden_size=64,          # rnn hidden unit
            num_layers=1,            # number of rnn layer
            batch_first=True,        # input & output will has batch size as
                                       1s dimension. e.g. (batch, time_step,
                                       input_size)
        )

        self.out = nn.Linear(64, 10)

    def forward(self, x):
        # x shape (batch, time_step, input_size)
        # r_out shape (batch, time_step, output_size)
        # h_n shape (n_layers, batch, hidden_size)
        # h_c shape (n_layers, batch, hidden_size)
        r_out, (h_n, h_c) = self.rnn(x, None)   # None represents zero
                                                  initial hidden state

        # choose r_out at the last time step
        out = self.out(r_out[:, -1, :])
        return out
```

In the preceding RNN class, you are training an LSTM network, which is proven effective for holding memory for a long time, and thus helps in learning. If you use the nn.RNN() model, it hardly learns the parameters, because the vanilla implementation of RNN cannot hold or remember the information for a long period of time. In the LSTM network, the image width is considered the input size, the hidden size is decided as the number of neurons in the hidden layer, and num_layers shows the number of RNN layers in the network.

The RNN module, within the LSTM module, produces the output as a vector size of 64×10 because the output layer has digits to be classified as 0 to 9. The last forward function shows how to proceed with forward propagation in an RNN network.

The following script shows how the LSTM model is processed under the RNN class. In the LSTM function, you pass the input length as 28 and the number of neurons in the hidden layer as 64, and from the hidden 64 neurons to the output 10 neurons.

```
rnn = RNN()
print(rnn)
RNN(
  (rnn): LSTM(28, 64, batch_first=True)
  (out): Linear(in_features-64, out_features=10, bias=True)
)

optimizer = torch.optim.Adam(rnn.parameters(), lr-LR)   # optimize all RNN
                                                          parameters
loss_func = nn.CrossEntropyLoss()                        # the target label
                                                          is not one-hotted
```

To optimize all RNN parameters, you use the Adam optimizer. Inside the function, you use the learning rate as well. The loss function used in this example is the cross-entropy loss function. You need to provide multiple epochs to get the best parameters.

In the following script, you print the training loss and the test accuracy. After one epoch, the test accuracy increases to 95% and the training loss reduces to 0.24.

```
# training and testing
for epoch in range(EPOCH):
    for step, (x, y) in enumerate(train_loader):       # gives batch data
        b_x = Variable(x.view(-1, 28, 28))             # reshape x to
                                                         (batch, time_
                                                         step, input_size)

        b_y = Variable(y)                              # batch y

        output = rnn(b_x)                              # rnn output
```

```
        loss = loss_func(output, b_y)                        # cross
                                                               entropy loss

        optimizer.zero_grad()                                # clear gradients
                                                               for this
                                                               training step

        loss.backward()                                      # backpropagation,
                                                               compute gradients

        optimizer.step()                                     # apply gradients

        if step % 50 == 0:
            test_output = rnn(test_x)                        # (samples, time_
                                                               step, input_size)

            pred_y = torch.max(test_output, 1)[1].data.numpy().squeeze()
            accuracy = sum(pred_y == test_y) / float(test_y.size)
            print('Epoch: ', epoch, '| train loss: %.4f' % loss.data,
            '| test accuracy: %.2f' % accuracy)
```

```
Epoch:  0 | train loss: 2.3088 | test accuracy: 0.09
Epoch:  0 | train loss: 1.3125 | test accuracy: 0.59
Epoch:  0 | train loss: 0.8936 | test accuracy: 0.71
Epoch:  0 | train loss: 0.4285 | test accuracy: 0.83
Epoch:  0 | train loss: 0.2509 | test accuracy: 0.87
Epoch:  0 | train loss: 0.3429 | test accuracy: 0.90
Epoch:  0 | train loss: 0.3704 | test accuracy: 0.86
Epoch:  0 | train loss: 0.4593 | test accuracy: 0.91
Epoch:  0 | train loss: 0.0794 | test accuracy: 0.94
Epoch:  0 | train loss: 0.0768 | test accuracy: 0.93
Epoch:  0 | train loss: 0.1809 | test accuracy: 0.94
Epoch:  0 | train loss: 0.2297 | test accuracy: 0.94
Epoch:  0 | train loss: 0.2210 | test accuracy: 0.95
Epoch:  0 | train loss: 0.2509 | test accuracy: 0.94
Epoch:  0 | train loss: 0.0828 | test accuracy: 0.94
Epoch:  0 | train loss: 0.2879 | test accuracy: 0.95
Epoch:  0 | train loss: 0.0908 | test accuracy: 0.94
Epoch:  0 | train loss: 0.1554 | test accuracy: 0.94
Epoch:  0 | train loss: 0.1557 | test accuracy: 0.96
```

Once the model is trained, the next step is to make predictions using the RNN model. Then you compare the actual vs. real output to assess how the model is performing.

```
# print 10 predictions from test data
test_output = rnn(test_x[:10].view(-1, 28, 28))
pred_y = torch.max(test_output, 1)[1].data.numpy().squeeze()
print(pred_y, 'prediction number')
print(test_y[:10], 'real number')

[7 2 1 0 4 1 4 9 5 9] prediction number
[7 2 1 0 4 1 4 9 5 9] real number
```

Recipe 3-9. Implementing a RNN for Regression Problems

Problem

How do you set up a recurrent neural network for regression-based problems?

Solution

The regression model requires a target function and a feature set, and then a function to establish the relationship between the input and the output. In this example, you are going to use the recurrent neural network for a regression task. Regression problems seem to be very simple; they do work best but are limited to data that shows clear linear relationships. They are quite complex when predicting nonlinear relationships between the input and the output.

How It Works

Let's look at the following example that shows a nonlinear cyclical pattern between input and output data. In the previous recipe, you looked at an example of a RNN in general for classification-related problems, where it predicted the class of the input image. In regression, however, the architecture of a RNN changes because the objective is to predict the real valued output. The output layer would have one neuron in regression-related problems.

87

```
import torch
from torch import nn
from torch.autograd import Variable
import numpy as np
import matplotlib.pyplot as plt
%matplotlib inline

torch.manual_seed(1)    # reproducible

# Hyper Parameters
TIME_STEP = 10        # rnn time step
INPUT_SIZE = 1        # rnn input size
LR = 0.02             # learning rate
```

The RNN time step implies that the last 10 values predict the current value, and the rolling happens after that.

The following script shows some sample series in which the target cos function is approximated by the sin function. See Figure 3-6.

```
# show data
steps = np.linspace(0, np.pi*2, 100, dtype=np.float32)
x_np = np.sin(steps)     # float32 for converting torch FloatTensor
y_np = np.cos(steps)
plt.plot(steps, y_np, 'r-', label='target (cos)')
plt.plot(steps, x_np, 'b-', label='input (sin)')
plt.legend(loc='best')
plt.show()
```

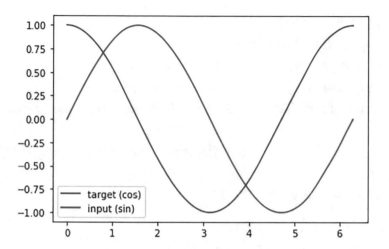

Figure 3-6.

Recipe 3-10. Using PyTorch's Built-In Functions

Problem

How do you set up an RNN module and call the RNN function using PyTorch?

Solution

By using the built-in function available in the neural network module, you can implement an RNN model.

How It Works

Let's look at the following example. The neural network module in the PyTorch library contains the RNN function. In the following script, you use the input matrix size, the number of neurons in the hidden layer, and the number of hidden layers in the network.

```
class RNN(nn.Module):
    def __init__(self):
        super(RNN, self).__init__()
```

```python
        self.rnn = nn.RNN(
            input_size=INPUT_SIZE,
            hidden_size=32,        # rnn hidden unit
            num_layers=1,          # number of rnn layer
            batch_first=True,      # input & output will has batch size as 1s
                                      dimension. e.g.
                                   # (batch, time_step, input_size)
        )
        self.out = nn.Linear(32, 1)

    def forward(self, x, h_state):
        # x (batch, time_step, input_size)
        # h_state (n_layers, batch, hidden_size)
        # r_out (batch, time_step, hidden_size)
        r_out, h_state = self.rnn(x, h_state)

        outs = []    # save all predictions
        for time_step in range(r_out.size(1)):    # calculate output for
                                                      each time step
            outs.append(self.out(r_out[:, time_step, :]))
        return torch.stack(outs, dim=1), h_state
```

After creating the RNN class function, you need to provide the optimization function, which is Adam, and this time, the loss function is the mean square loss function. Since the objective is to make predictions of a continuous variable, you use the MSELoss function in the optimization layer.

```python
rnn = RNN()
print(rnn)
RNN(
  (rnn): RNN(1, 32, batch_first=True)
  (out): Linear(in_features=32, out_features=1, bias=True)
)

optimizer = torch.optim.Adam(rnn.parameters(), lr=LR)    # optimize all cnn
                                                             parameters
loss_func = nn.MSELoss()
```

```
h_state = None        # for initial hidden state

plt.figure(1, figsize=(12, 5))
plt.ion()             # continuously plot

for step in range(60):
    start, end = step * np.pi, (step+1)*np.pi    # time range
    # use sin predicts cos
    steps = np.linspace(start, end, TIME_STEP, dtype=np.float32)
    x_np = np.sin(steps)     # float32 for converting torch FloatTensor
    y_np = np.cos(steps)

    x = Variable(torch.from_numpy(x_np[np.newaxis, :, np.newaxis]))
    # shape (batch, time_step, input_size)
    y = Variable(torch.from_numpy(y_np[np.newaxis, :, np.newaxis]))

    prediction, h_state = rnn(x, h_state)    # rnn output
    # !! next step is important !!
    h_state = Variable(h_state.data)
    # repack the hidden state, break the connection from last iteration

    loss = loss_func(prediction, y)          # cross entropy loss
    optimizer.zero_grad()                    # clear gradients for this
                                             #   training step
    loss.backward()                          # backpropagation, compute
                                             #   gradients
    optimizer.step()                         # apply gradients

    # plotting
    plt.plot(steps, y_np.flatten(), 'r-')
    plt.plot(steps, prediction.data.numpy().flatten(), 'b-')
    plt.draw(); plt.pause(0.05)
```

Now you iterate over 60 steps to predict the cos function generated from the sample space and have it predicted by a sin function. The iterations take the learning rate defined as before and backpropagate the error to reduce the MSE and improve the prediction. See Figure 3-7.

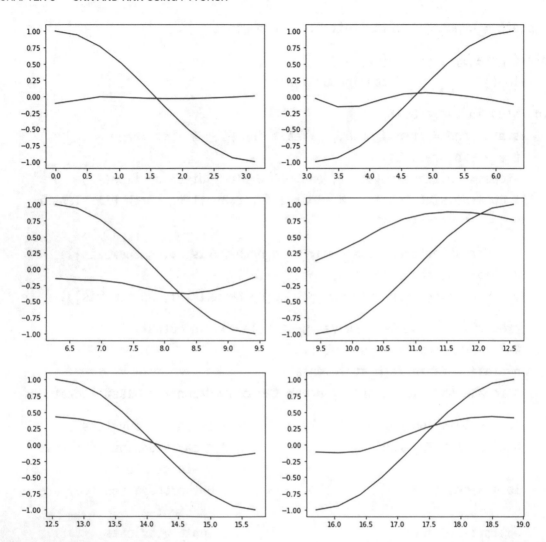

Figure 3-7.

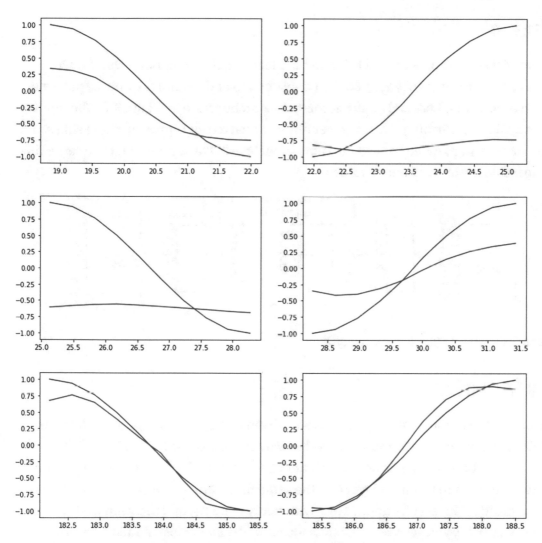

Figure 3-7. *(continued)*

Recipe 3-11. Working with Autoencoders
Problem

How do you perform clustering using the autoencoders function?

Solution

Unsupervised learning is a branch of machine learning that does not have a target column or the output is not defined. You only need to understand the unique patterns existing in the data. Let's look at the autoencoder architecture in Figure 3-8. The input feature space is transformed into a lower dimensional tensor representation using a hidden layer and mapped back to the same input space. The layer that is precisely in the middle holds the autoencoder's values.

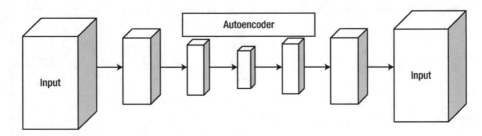

Figure 3-8. *Autoencoder architecture*

How It Works

Let's look at the following example. The torchvision library contains popular datasets, model architectures, and frameworks. Autoencoder is a process of identifying latent features from the dataset; it is used for classification, prediction, and clustering. If you put the input data in the input layer and the same dataset in the output layer, and then you add multiple layers of hidden layers with many neurons, and then you pass through a series of epochs, you get a set of latent features in the innermost hidden layer. The weights or parameters in the central hidden layer are known as the *autoencoder layer*.

```
import torch
import torch.nn as nn
from torch.autograd import Variable
import torch.utils.data as Data
import torchvision
import matplotlib.pyplot as plt
from mpl_toolkits.mplot3d import Axes3D
from matplotlib import cm
import numpy as np
%matplotlib inline
```

```
torch.manual_seed(1)     # reproducible

# Hyper Parameters
EPOCH = 10
BATCH_SIZE = 64
LR = 0.005           # learning rate
DOWNLOAD_MNIST = False
N_TEST_IMG = 5
```

You again use the MNIST dataset to experiment with autoencoder functionality. This time you are taking 10 epochs, a batch size 64 to be passed to the network, a learning rate of 0.005, and 5 images for testing.

```
# Mnist digits dataset
train_data = torchvision.datasets.MNIST(
    root='./mnist/',
    train=True,
    # this is training data
    transform=torchvision.transforms.ToTensor(),
    # Converts a PIL.Image or numpy.ndarray to

    # torch.FloatTensor of shape (C x H x W) and normalize in the range
      [0.0, 1.0]
    download=DOWNLOAD_MNIST,
    # download it if you don't have it
)
```

Figure 3-9 shows the dataset uploaded from the torchvision library and displayed as an image.

```
# plot one example
print(train_data.train_data.size())      # (60000, 28, 28)
print(train_data.train_labels.size())    # (60000)
plt.imshow(train_data.train_data[2].numpy(), cmap='gray')
plt.title('%i' % train_data.train_labels[2])
plt.show()
```

```
torch.Size([60000, 28, 28])
torch.Size([60000])
```

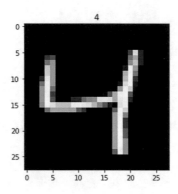

Figure 3-9.

```
# Data Loader for easy mini-batch return in training, the image batch shape
will be (50, 1, 28, 28)
train_loader = Data.DataLoader(dataset=train_data, batch_size=BATCH_SIZE,
shuffle=True)

class AutoEncoder(nn.Module):
    def __init__(self):
        super(AutoEncoder, self).__init__()

        self.encoder = nn.Sequential(
            nn.Linear(28*28, 128),
            nn.Tanh(),
            nn.Linear(128, 64),
            nn.Tanh(),
            nn.Linear(64, 12),
            nn.Tanh(),
            nn.Linear(12, 3),    # compress to 3 features which can be
                                   visualized in plt
        )
        self.decoder = nn.Sequential(
            nn.Linear(3, 12),
            nn.Tanh(),
            nn.Linear(12, 64),
```

```
            nn.Tanh(),
            nn.Linear(64, 128),
            nn.Tanh(),
            nn.Linear(128, 28*28),
            nn.Sigmoid(),        # compress to a range (0, 1)
        )

    def forward(self, x):
        encoded = self.encoder(x)
        decoded = self.decoder(encoded)
        return encoded, decoded
```

Let's discuss the autoencoder architecture. The input has 784 features. It has a height of 28 and a width of 28. You pass the 784 neurons from the input layer to the first hidden layer, which has 128 neurons in it. Then you apply the hyperbolic tangent function to pass the information to the next hidden layer. The second hidden layer contains 128 input neurons and transforms it into 64 neurons. In the third hidden layer, you apply the hyperbolic tangent function to pass the information to the next hidden layer. The innermost layer contains three neurons, which are considered as three features, which is the end of the encoder layer. Then the decoder function expands the layer back to the 784 features in the output layer.

```
autoencoder = AutoEncoder()
print(autoencoder)

optimizer = torch.optim.Adam(autoencoder.parameters(), lr=LR)
loss_func = nn.MSELoss()

# original data (first row) for viewing
view_data = Variable(train_data.train_data[:N_TEST_IMG].view(-1, 28*28).
type(torch.FloatTensor)/255.)

AutoEncoder(
  (encoder): Sequential(
    (0): Linear(in_features=784, out_features=128, bias=True)
    (1): Tanh()
    (2): Linear(in_features=128, out_features=64, bias=True)
    (3): Tanh()
```

```
    (4): Linear(in_features=64, out_features=12, bias=True)
    (5): Tanh()
    (6): Linear(in_features=12, out_features=3, bias=True)
  )
  (decoder): Sequential(
    (0): Linear(in_features=3, out_features=12, bias=True)
    (1): Tanh()
    (2): Linear(in_features=12, out_features=64, bias=True)
    (3): Tanh()
    (4): Linear(in_features=64, out_features=128, bias=True)
    (5): Tanh()
    (6): Linear(in_features=128, out_features=784, bias=True)
    (7): Sigmoid()
  )
)
```

Once you set the architecture, then the normal process of making the loss function minimize corresponding to a learning rate and optimization function happens. The entire architecture passes through a series of epochs in order to reach the target output.

Recipe 3-12. Fine-Tuning Results Using Autoencoder

Problem

How do you set up iterations to fine-tune the results?

Solution

Conceptually, an autoencoder works the same as the clustering model. In unsupervised learning, the machine learns patterns from data and generalizes it to the new dataset. The learning happens by taking a set of input features. Autoencoder functions are also used for feature engineering.

How It Works

Let's look at the following example. The same MNIST dataset is used as an example, and the objective is to understand the role of the epoch in achieving a better autoencoder layer. You increase the epoch size to reduce errors to a minimum; however, in practice, increasing the epoch has many challenges, including memory constraints. See Figure 3-10.

```python
for epoch in range(EPOCH):
    for step, (x, y) in enumerate(train_loader):
        b_x = Variable(x.view(-1, 28*28))    # batch x, shape (batch, 28*28)
        b_y = Variable(x.view(-1, 28*28))    # batch y, shape (batch, 28*28)
        b_label = Variable(y)                # batch label

        encoded, decoded = autoencoder(b_x)

        loss = loss_func(decoded, b_y)       # mean square error
        optimizer.zero_grad()                # clear gradients for this
                                             # training step
        loss.backward()                      # backpropagation, compute
                                             # gradients
        optimizer.step()                     # apply gradients

        if step % 500 == 0 and epoch in [0, 5, EPOCH-1]:
            print('Epoch: ', epoch, '| train loss: %.4f' % loss.data)

            # plotting decoded image (second row)
            _, decoded_data = autoencoder(view_data)

            # initialize figure
            f, a = plt.subplots(2, N_TEST_IMG, figsize=(5, 2))

            for i in range(N_TEST_IMG):
                a[0][i].imshow(np.reshape(view_data.data.numpy()[i],
                                          (28, 28)), cmap='gray');
                a[0][i].set_xticks(()); a[0][i].set_yticks(())

            for i in range(N_TEST_IMG):
                a[1][i].clear()
```

```
a[1][i].imshow(np.reshape(decoded_data.data.numpy()[i],
                          (28, 28)), cmap='gray')
a[1][i].set_xticks(()); a[1][i].set_yticks(())
plt.show(); #plt.pause(0.05)
```

Epoch: 0 | train loss: 0.2213

Epoch: 0 | train loss: 0.0678

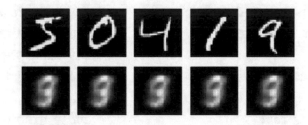

Epoch: 5 | train loss: 0.0375

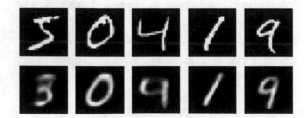

Figure 3-10.

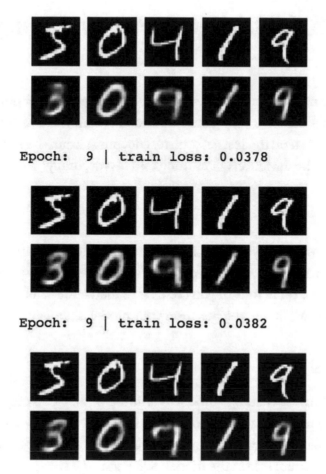

Figure 3-10. *(continued)*

By using the encoder function, you can represent the input features into a set of latent features. By using the decoder function, however, you can reconstruct the image. Then you can match how image reconstruction is done by using the autoencoder functions. From the preceding set of graphs, it is clear that as you increase the epoch, the image recognition becomes apparent.

```
torch.manual_seed(1)     # reproducible
```

Recipe 3-13. Restricting Model Overfitting

Problem

When you fit many neurons and layers to predict the target class or output variable, the function usually overfits the training dataset. Because of model overfitting, you cannot make a good prediction on the test set. The test accuracy is not the same as training accuracy. There will be deviations in training and test accuracy.

Solution

To restrict model overfitting, you consciously introduce dropout rate, which means randomly delete (let's say) 10% or 20% of the weights in the network and check the model accuracy at the same time. If you are able to match the same model accuracy after deleting the 10% or 20% of the weights, then your model is good.

How It Works

Let's look at the following example. Model overfitting occurs when the trained model does not generalize to other test case scenarios. It is identified when the training accuracy deviates significantly from the test accuracy. To avoid model overfitting, you can introduce the dropout rate in the model. See Figure 3-11.

```
import torch
from torch.autograd import Variable
import matplotlib.pyplot as plt
%matplotlib inline

torch.manual_seed(1)    # reproducible
N_SAMPLES = 20
N_HIDDEN = 300

# training data
x = torch.unsqueeze(torch.linspace(-1, 1, N_SAMPLES), 1)
y = x + 0.3*torch.normal(torch.zeros(N_SAMPLES, 1), torch.ones
(N_SAMPLES, 1))
x, y = Variable(x), Variable(y)
```

```
# test data
test_x = torch.unsqueeze(torch.linspace(-1, 1, N_SAMPLES), 1)
test_y = test_x + 0.3*torch.normal(torch.zeros(N_SAMPLES, 1), torch.ones
(N_SAMPLES, 1))
test_x, test_y = Variable(test_x), Variable(test_y )

# show data
plt.scatter(x.data.numpy(), y.data.numpy(), c='magenta', s=50, alpha=0.5,
label='train')
plt.scatter(test_x.data.numpy(), test_y.data.numpy(), c='cyan', s=50,
alpha=0.5, label='test')
plt.legend(loc='upper left')
plt.ylim((-2.5, 2.5))
plt.show()
```

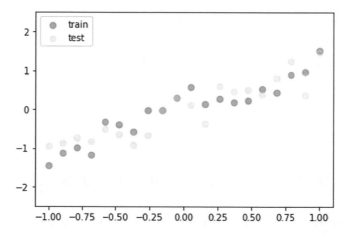

Figure 3-11.

The dropout rate introduction to the hidden layer ensures that weights less than the threshold defined are removed from the architecture. A typical threshold for an application's dropout rate is 20% to 50%. A 20% dropout rate implies a smaller degree of penalization; however, the 50% threshold implies heavy penalization of the model weights.

In the following script, you apply a 50% dropout rate to drop the weights from the model. You apply the dropout rate twice.

```python
net_overfitting = torch.nn.Sequential(
    torch.nn.Linear(1, N_HIDDEN),
    torch.nn.ReLU(),
    torch.nn.Linear(N_HIDDEN, N_HIDDEN),
    torch.nn.ReLU(),
    torch.nn.Linear(N_HIDDEN, 1),
)

net_dropped = torch.nn.Sequential(
    torch.nn.Linear(1, N_HIDDEN),
    torch.nn.Dropout(0.5),   # drop 50% of the neuron
    torch.nn.ReLU(),
    torch.nn.Linear(N_HIDDEN, N_HIDDEN),
    torch.nn.Dropout(0.5),   # drop 50% of the neuron
    torch.nn.ReLU(),
    torch.nn.Linear(N_HIDDEN, 1),
)
print(net_overfitting)  # net architecture
print(net_dropped)

Sequential(
  (0): Linear(in_features=1, out_features=300, bias=True)
  (1): ReLU()
  (2): Linear(in_features=300, out_features=300, bias=True)
  (3): ReLU()
  (4): Linear(in_features=300, out_features=1, bias=True)
)
Sequential(
  (0): Linear(in_features=1, out_features=300, bias=True)
  (1): Dropout(p=0.5, inplace=False)
  (2): ReLU()
  (3): Linear(in_features=300, out_features=300, bias=True)
  (4): Dropout(p=0.5, inplace=False)
  (5): ReLU()
  (6): Linear(in_features=300, out_features=1, bias=True)
)
```

```
optimizer_ofit = torch.optim.Adam(net_overfitting.parameters(), lr=0.01)
optimizer_drop = torch.optim.Adam(net_dropped.parameters(), lr=0.01)
loss_func = torch.nn.MSELoss()
```

The selection of the right dropout rate requires a fair idea about the business and domain.

Recipe 3-14. Visualizing the Model Overfit Problem

Evaluate deep learning model overfitting.

Solution

Change the model hyperparameters and iteratively see if the model is overfitting data or not.

How It Works

The previous recipe covered two types of neural networks: overfitting and dropout rate. When the model parameters estimated from the data come closer to the actual data for the training dataset, and the same models differs from the test set, it is a clear sign of model overfit. To restrict model overfit, you can introduce the dropout rate, which deletes a certain percentage of connections (as in weights from the network) to allow the trained model to come to the real data.

In the following script, the iterations are taken 500 times. The predicted values are generated from the base model, which shows overfitting, and from the dropout model, which shows the deletion of some weights. In the same fashion, you create the two loss functions, backpropagation, and implementation of the optimizer. See Figure 3-12.

```
for t in range(500):
    pred_ofit = net_overfitting(x)
    pred_drop = net_dropped(x)
    loss_ofit = loss_func(pred_ofit, y)
    loss_drop = loss_func(pred_drop, y)
```

```
optimizer_ofit.zero_grad()
optimizer_drop.zero_grad()
loss_ofit.backward()
loss_drop.backward()
optimizer_ofit.step()
optimizer_drop.step()

if t % 100 == 0:
    # change to eval mode in order to fix drop out effect
    net_overfitting.eval()
    net_dropped.eval()  # parameters for dropout differ from train mode

    # plotting
    plt.cla()
    test_pred_ofit = net_overfitting(test_x)
    test_pred_drop = net_dropped(test_x)
    plt.scatter(x.data.numpy(), y.data.numpy(), c='magenta', s=50,
                alpha=0.3, label='train')
    plt.scatter(test_x.data.numpy(), test_y.data.numpy(),
    c='cyan', s=50,
                alpha=0.3, label='test')
    plt.plot(test_x.data.numpy(), test_pred_ofit.data.numpy(), 'r-',
            lw=3, label='overfitting')
    plt.plot(test_x.data.numpy(), test_pred_drop.data.numpy(), 'b--',
            lw=3, label='dropout(50%)')
    plt.text(0, -1.2, 'overfitting loss=%.4f' % loss_func(test_pred_
    ofit, test_y).data,
            fontdict={'size': 20, 'color':  'red'})
    plt.text(0, -1.5, 'dropout loss=%.4f' % loss_func(test_pred_drop,
    test_y).data,
            fontdict={'size': 20, 'color': 'blue'})
    plt.legend(loc='upper left'); plt.ylim((-2.5, 2.5));plt.pause(0.1)

    # change back to train mode
    net_overfitting.train()
    net_dropped.train()
    plt.show()
```

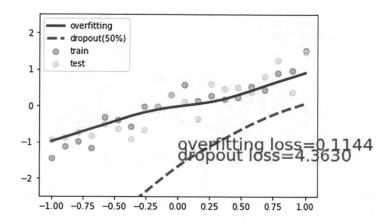

Figure 3-12.

The initial round of plotting includes the overfitting loss and dropout loss and how it is different from the actual training and test data points from the preceding graph. See Figure 3-13.

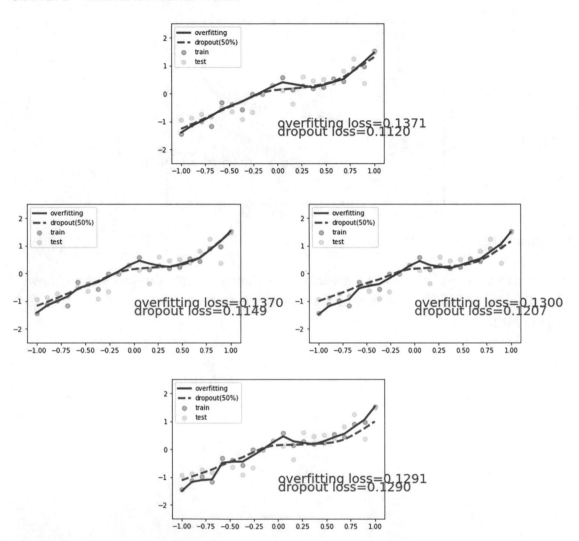

Figure 3-13.

After many iterations, the preceding graph was generated by using the two functions with the actual model and with the dropout rate. The takeaway from this graph is that actual training data may get closer to the overfit model; however, the dropout model fits the data really well.

Recipe 3-15. Initializing Weights in the Dropout Rate

Problem

How do you delete the weights in a network? Should you delete randomly or by using a distribution?

Solution

You should delete the weights in the dropout layer based on probability distribution, rather than randomly.

How It Works

In the previous recipe, three layers of a dropout rate were introduced: one after the first hidden layer and two after the second hidden layer. The probability percentage was 0.50, which meant randomly delete 50% of the weights. Sometimes, random selection of weights from the network deletes relevant weights, so an alternative idea is to delete the weights in the network generated from statistical distribution.

   The following script shows how to generate the weights from a uniform distribution and then you can use the set of weights in the network architecture.

```
import numpy as np
import torch

#From a uniform distribution
torch.Tensor(5, 3)
tensor([[2.6019e-33, 0.0000e+00, 3.7835e-44], [0.0000e+00, nan,
0.0000e+00], [1.3733e-14, 6.4069e+02, 4.3066e+21], [1.1824e+22, 4.3066e+21,
6.3828e+28], [3.8016e-39, 0.0000e+00, 1.5501e-37]])

#getting the shape of the tensor
torch.Tensor(5, 3).uniform_(-1, 1)
tensor([[ 0.8790, 0.7375, 0.1182], [ 0.3652, 0.1322, 0.8587], [ 0.3682,
-0.2907, 0.0051], [ 0.0886, -0.7588, -0.5371], [ 0.0085, 0.0812, -0.6360]])
```

```
#getting the shape of the tensor
x = torch.Tensor(5, 3).uniform_(-1, 1)
print(x.size())
torch.Size([5, 3])

#Creation from lists & numpy
z = torch.LongTensor([[1, 3], [2, 9]])
print(z.type())
# Cast to numpy ndarray
print(z.numpy().dtype)
torch.LongTensor
int64

# Data type inferred from numpy
print(torch.from_numpy(np.random.rand(5, 3)).type())
print(torch.from_numpy(np.random.rand(5, 3).astype(np.float32)).type())
torch.DoubleTensor
torch.FloatTensor
```

Recipe 3-16. Adding Math Operations

Problem

How do you set up the broadcasting function and optimize the convolution function?

Solution

The script snippet shows how to introduce batch normalization when setting up a convolutional neural network model and then further setting up a pooling layer.

How It Works

To introduce batch normalization in the convolutional layer of the neural network model, you need to perform tensor-based mathematical operations that are functionally different from other methods of computation.

```
#Simple mathematical operations
y = x * torch.randn(5, 3)
print(y)
tensor([[ 0.1587,  0.4137, -0.4801],
        [-0.2706,  0.0411, -0.8954],
        [ 0.3616, -0.0245, -0.3401],
        [-0.6478, -0.1207, -0.1698],
        [ 0.2107, -0.2128,  0.1017]])

y = x / torch.sqrt(torch.randn(5, 3) ** 2)
print(y)
tensor([[ 2.1697, -1.1561, -7.4875],
        [-0.5094,  0.4193, -4.4016],
        [ 0.4308,  0.0421,  0.6234],
        [ 2.3634,  2.1020, -0.2185],
        [ 4.8023,  0.4352,  0.1892]])

#Broadcasting
print (x.size())
y = x + torch.randn(5, 1)
print(y)
torch.Size([5, 3])
tensor([[ 1.0416, -0.1192, -0.1256],
        [ 0.0484,  1.5687,  0.0468],
        [ 0.1000, -0.4971,  0.3657],
        [ 0.3893,  0.5367, -0.2656],
        [ 2.1538,  1.9121,  2.0349]])

#Reshape
y = torch.randn(5, 10, 15)
print(y.size())
print(y.view(-1, 15).size())  # Same as doing y.view(50, 15)
print(y.view(-1, 15).unsqueeze(1).size()) # Adds a dimension at index 1.
print(y.view(-1, 15).unsqueeze(1).squeeze().size())
print()
print(y.transpose(0, 1).size())
print(y.transpose(1, 2).size())
```

```
print(y.transpose(0, 1).transpose(1, 2).size())
print(y.permute(1, 2, 0).size())
torch.Size([5, 10, 15])
torch.Size([50, 15])
torch.Size([50, 1, 15])
torch.Size([50, 15])

torch.Size([10, 5, 15])
torch.Size([5, 15, 10])
torch.Size([10, 15, 5])
torch.Size([10, 15, 5])

#Repeat
print(y.view(-1, 15).unsqueeze(1).expand(50, 100, 15).size())
print(y.view(-1, 15).unsqueeze(1).expand_as(torch.randn(50, 100,
15)).size())
torch.Size([50, 100, 15])
torch.Size([50, 100, 15])

#Concatenate tensors
# 2 is the dimension over which the tensors are concatenated
print(torch.cat([y, y], 2).size())
# stack concatenates the sequence of tensors along a new dimension.
print(torch.stack([y, y], 0).size())

torch.Size([5, 10, 30])
torch.Size([2, 5, 10, 15])

#Advanced Indexing
y = torch.randn(2, 3, 4)
print(y[[1, 0, 1, 1]].size())

# PyTorch doesn't support negative strides yet so ::-1 does not work.
rev_idx = torch.arange(1, -1, -1).long()
print(y[rev_idx].size())

torch.Size([4, 3, 4])
torch.Size([2, 3, 4])
```

The following script shows how the batch normalization using a 2D layer is resolved before entering into the 2D max pooling layer:

```
#Convolution, BatchNorm & Pooling Layers
x = Variable(torch.randn(10, 3, 28, 28))

conv = nn.Conv2d(in_channels=3, out_channels=32, kernel_size=(3, 3), stride=1,
                  padding=1, bias=True)
bn = nn.BatchNorm2d(num_features=32)
pool = nn.MaxPool2d(kernel_size=(2, 2), stride=2)

output_conv = bn(conv(x))
outpout_pool = pool(conv(x))

print('Conv output size : ', output_conv.size())
print('Pool output size : ', outpout_pool.size())
Conv output size :  torch.Size([10, 32, 28, 28])
Pool output size :  torch.Size([10, 32, 14, 14])
```

Recipe 3-17. Embedding Layers in RNN

Problem

The recurrent neural network is used mostly for text processing. An embedded feature offers more accuracy on a standard RNN model than raw features. How do you create embedded features in an RNN?

Solution

The first step is to create an embedding layer, which is a fixed dictionary and fixed-size lookup table, and then introduce the dropout rate after than create gated recurrent unit.

How It Works

When textual data comes in as a sequence, the information is processed in a sequential way; for example, when we describe something, we use a set of words in sequence to convey the meaning. If we use the individual words as vectors to represent the data, the resulting dataset would be very sparse. But if we use a phrase-based approach or a combination of words to represent as feature vector, then the vectors become a dense layer. Dense vector layers are called *word embeddings,* as the embedding layer conveys a context or meaning as the result. It is definitely better than the bag-of-words approach.

```
#Recurrent, Embedding & Dropout Layers
inputs = [[1, 2, 3], [1, 0, 4], [1, 2, 4], [1, 4, 0], [1, 3, 3]]
x = Variable(torch.LongTensor(inputs))

embedding = nn.Embedding(num_embeddings=5, embedding_dim=20, padding_idx=1)
drop = nn.Dropout(p=0.5)
gru = nn.GRU(input_size=20, hidden_size=50, num_layers=2, batch_first=True,
            bidirectional=True, dropout=0.3)

emb = drop(embedding(x))
gru_h, gru_h_t = gru(emb)

print('Embedding size : ', emb.size())
print('GRU hidden states size : ', gru_h.size())
print('GRU last hidden state size : ', gru_h_t.size())
Embedding size :  torch.Size([5, 3, 20])
GRU hidden states size :  torch.Size([5, 3, 100])
GRU last hidden state size :  torch.Size([4, 5, 50])

#The functional API provides users a way to use these classes in a
functional way.
import torch.nn.functional as F
x = Variable(torch.randn(10, 3, 28, 28))
filters = Variable(torch.randn(32, 3, 3, 3))
conv_out = F.relu(F.dropout(F.conv2d(input=x, weight=filters, padding=1),
                        p=0.5, training=True))

print('Conv output size : ', conv_out.size())
Conv output size :  torch.Size([10, 32, 28, 28])
```

Conclusion

This chapter covered using the PyTorch API, creating a simple neural network mode, and optimizing the parameters by changing the hyperparameters (i.e., learning rate, epochs, gradients drop). You looked at recipes on how to create a convolutional neural network and a recurrent neural network, and you introduced the dropout rate in these networks to control model overfitting.

You used small tensors to follow what exactly goes on behind the scenes with calculations and so forth. You only need to define the problem statement, create features, and apply the recipe to get results. In the next chapter, you will implement many more examples with PyTorch.

## CHAPTER 4

Introduction to Neural Networks Using PyTorch

Deep neural network–based models are gradually becoming the backbone for artificial intelligence and machine learning implementations. The future of data mining will be governed by the usage of artificial neural network–based advanced modeling techniques. One obvious question is why neural networks are only now gaining so much importance, because they were invented in 1950s.

Borrowed from the computer science domain, neural networks can be defined as parallel information processing systems where all input relates to each other, like neurons in the human brain, to transmit information so that activities like facial recognition and image recognition can be performed. In this chapter, you will learn about the application of neural network-based methods on various data mining tasks, such as classification, regression, forecasting, and feature reduction. An *artificial neural network* (ANN) functions similarly to the way the human brain functions, in which billions of neurons link to each other for information processing and insight generation.

Recipe 4-1. Working with Activation Functions

Problem

What are activation functions and how do they work in real projects? How do you implement an activation function using PyTorch?

Solution

An activation function is a mathematical formula that transforms a vector available in a binary, float, or integer format to another format based on the type of mathematical transformation function. The neurons are present in different layers—input, hidden, and

117

© Pradeepta Mishra 2023
P. Mishra, *PyTorch Recipes*, https://doi.org/10.1007/978-1-4842-8925-9_4

output, which are interconnected through a mathematical function called an *activation function*. There are different variants of activation functions, which are explained next. Understanding the activation function helps in accurately implementing a neural network model.

How It Works

All activation functions that are part of a neural network model can be broadly classified as linear functions and nonlinear functions. The PyTorch `torch.nn` module creates any type of a neural network model. Let's look at some examples of the deployment of activation functions using PyTorch and the `torch.nn` module.

The core differences between PyTorch and TensorFlow are the way a computational graph is defined, the way the two frameworks perform calculations, and the amount of flexibility you have in changing the script and introducing other Python-based libraries in it. In TensorFlow, you must define the variables and placeholders before you initialize the model. You must also keep track of objects you need later, and for that you need a placeholder. In TensorFlow, you need to define the model first and then compile and run it; however, in PyTorch, you can define the model as you go—you don't have to keep placeholders in the code. That's why the PyTorch framework is dynamic.

Linear Function

A linear function is a simple function typically used to transfer information from the demapping layer to the output layer. You use a linear function in places where variations in data are lower. In a deep learning model, practitioners typically use a linear function in the last hidden layer to the output layer. In a linear function, the output is always confined to a specific range; because of that, it is used in the last hidden layer in a deep learning model, or in linear regression–based tasks, or in a deep learning model where the task is to predict the outcome from the input dataset. Here is the formula:

$$y = \alpha + \beta x$$

Bilinear Function

A bilinear function is a simple function typically used to transfer information. It applies a bilinear transformation to incoming data.

$$y = x_1 * A * x_2 + b$$

```python
from __future__ import print_function
import torch
import numpy as np
import torch.optim
import torch.nn as nn
import torch.optim as optim
import torch.nn.init as init
import torch.nn.functional as F
from torch.autograd import Variable

import warnings
warnings.filterwarnings("ignore", category=FutureWarning)

#torch.nn: - Neural networks can be constructed using the torch.nn package.

x = Variable(torch.randn(100, 10))
y = Variable(torch.randn(100, 30))

linear = nn.Linear(in_features=10, out_features=5, bias=True)
output_linear = linear(x)
print('Output size : ', output_linear.size())

bilinear = nn.Bilinear(in1_features=10, in2_features=30, out_features=5,
bias=True)
output_bilinear = bilinear(x, y)
print('Output size : ', output_bilinear.size())
Output size :  torch.Size([100, 5])
Output size :  torch.Size([100, 5])
```

Sigmoid Function

A sigmoid function is frequently used by professionals in data mining and analytics because it is easier to explain and implement. It is a nonlinear function. When you pass weights from the input layer to the hidden layer in a neural network, you want your model to capture all sorts of nonlinearity present in the data; hence, using the sigmoid function in the hidden layers of a neural network is recommended. The nonlinear functions help with generalizing the dataset. It is easier to compute the gradient of a function using a nonlinear function.

The sigmoid function is a specific nonlinear activation function. The sigmoid function output is always confined within 0 and 1; therefore, it is mostly used in performing classification-based tasks. One of the limitations of the sigmoid function is that it may get stuck in local minima. An advantage is that it provides probability of belonging to the class. Here is its equation:

$$f(x) = \frac{1}{1 + e^{-\beta x}}$$

```
x = Variable(torch.randn(100, 10))
y = Variable(torch.randn(100, 30))

sig = nn.Sigmoid()
output_sig = sig(x)
output_sigy = sig(y)
print('Output size : ', output_sig.size())
print('Output size : ', output_sigy.size())

Output size :  torch.Size([100, 10])
Output size :  torch.Size([100, 30])

print(x[0])
print(output_sig[0])
tensor([-1.5454,  0.3599,  2.2720,  0.7115,  0.5296,  0.6176,  1.8854,  0.4854,
        -0.3893,  0.8369])
tensor([0.1758, 0.5890, 0.9065, 0.6707, 0.6294, 0.6497, 0.8682, 0.6190,
0.4039, 0.6978])
```

Hyperbolic Tangent Function

A hyperbolic tangent function is another variant of a transformation function. It is used to transform information from the mapping layer to the hidden layer. It is typically used between the hidden layers of a neural network model. The range of the tanh function is between –1 and +1.

$$\tanh(x) = \frac{e^x - e^{-x}}{e^x + e^{-x}}$$

```
x = Variable(torch.randn(100, 10))
y = Variable(torch.randn(100, 30))

func = nn.Tanh()
output_x = func(x)
output y = func(y)
print('Output size : ', output_x.size())
print('Output size : ', output_y.size())

Output size :  torch.Size([100, 10])
Output size :  torch.Size([100, 30])

print(x[0])
print(output_x[0])
print(y[0])
print(output_y[0])

tensor([ 1.6056,  0.1092,  0.2044,  1.0537, -0.8658, -0.9111, -1.1586, -1.7745,
        -0.8922, -2.3219])
tensor([ 0.9225,  0.1087,  0.2016,  0.7832, -0.6992, -0.7217, -0.8206, -0.9441,
        -0.7125, -0.9809])
tensor([ 0.2153,  1.3900,  0.4259, -0.3347, -1.2087, -0.1930,  0.1645, -1.5867,
        -0.1752,  0.3863,  0.6141,  1.6769, -0.8080,  0.3790, -0.7446,  0.1795,
        -1.5132,  0.8282,  1.6872,  0.7207, -0.6874,  0.0136,  0.3600,  1.9525,
        -0.1363, -0.2002,  0.4026, -0.1413,  2.2343,  1.0469])
tensor([ 0.2121,  0.8832,  0.4019, -0.3228, -0.8363, -0.1907,  0.1631, -0.9196,
        -0.1735,  0.3682,  0.5470,  0.9325, -0.6685,  0.3619, -0.6319,  0.1776,
```

-0.9075,   0.6795,   0.9338,   0.6173, -0.5963,   0.0136,   0.3452,

0.9605,

-0.1355, -0.1976,   0.3821, -0.1404,   0.9773,   0.7806])

Log Sigmoid Transfer Function

The following formula explains the log sigmoid transfer function, which is used in mapping the input layer to the hidden layer. If the data is not binary, and it is a float type with a lot of outliers (as in large numeric values present in the input feature), you should use the log sigmoid transfer function.

$$f(x) = \log\left(\frac{1}{1+e^{-\beta x}}\right)$$

```
x = Variable(torch.randn(100, 10))
y = Variable(torch.randn(100, 30))

func = nn.LogSigmoid()
output_x = func(x)
output_y = func(y)
print('Output size : ', output_x.size())
print('Output size : ', output_y.size())

Output size :   torch.Size([100, 10])
Output size :   torch.Size([100, 30])

print(x[0])
print(output_x[0])
print(y[0])
print(output_y[0])

tensor([-0.9983, -0.2337,  0.7794,  1.0399, -1.4705, -1.4177, -0.2531,
        -1.0391,
        -1.1570, -0.5105])
tensor([-1.3120, -0.8168, -0.3775, -0.3027, -1.6773, -1.6346, -0.8277, -1.3420,
        -1.4304, -0.9806])
tensor([-0.3758, -1.1889,  0.7846,  0.8277,  0.1351,  0.2677, -0.2810, -1.1610,
```

```
        -0.6973,  -0.1106,   0.6361,   1.4497,  -0.6007,  -0.1102,   0.8876,  -0.1440,
        -0.2914,  -0.0144,   1.4152,   2.1429,   0.8828,   0.9561,  -0.1876,   1.1487,
         0.6150,  -0.1044,   1.3075,  -0.1601,  -0.4018,  -1.2599])
tensor([-0.8986, -1.4547,  -0.3759,  -0.3626,  -0.6279,  -0.5683,  -0.8435,  -1.4335,
        -1.1014,  -0.7500,  -0.4249,  -0.2108,  -1.0379,  -0.7498,  -0.3448,  -0.7677,
        -0.8494,  -0.7004,  -0.2174,  -0.1109,  -0.3462,  -0.3253,  -0.7913,  -0.2754,
        -0.4322,  -0.7467,  -0.2394,  -0.7764,  -0.9141,  -1.5096])
```

ReLU Function

The rectified linear unit (ReLU) is another activation function. It is used in transferring information from the input layer to the output layer. ReLU is mostly used in a convolutional neural network model. The range in which this activation function operates is from 0 to infinity. It is mostly used between different hidden layers in a neural network model.

```
X = Variable(torch.randn(100, 10))
y = Variable(torch.randn(100, 30))

func = nn.ReLU()
output_x = func(x)
output_y = func(y)
print('Output size : ', output_x.size())
print('Output size : ', output_y.size())

Output size :  torch.Size([100, 10])
Output size :  torch.Size([100, 30])

print(x[0])
print(output_x[0])
print(y[0])
print(output_y[0])

tensor([-0.6479, -0.8856,  0.5144, -0.5064,  0.3280, -1.8378,  0.5670,
0.9095,  -2.6267, -1.0119])
tensor([0.0000, 0.0000, 0.5144, 0.0000, 0.3280, 0.0000, 0.5670, 0.9095,
        0.0000, 0.0000])
```

```
tensor([-1.4458,   0.8328,   0.6534,   2.0404,   0.9053, -0.2829,
        -0.5712,   0.0323,
         0.9757, -1.5787,   1.9665,   1.0276, -1.0536,   0.0588,   0.5085,   0.1956,
        -0.4490, -0.8927,   0.0128, -0.5971, -0.0677,   0.0101,   0.9477,   1.1218,
        -1.0648, -0.8439,   0.3422,   0.6930, -0.4311, -1.2920])
tensor([0.0000, 0.8328, 0.6534, 2.0404, 0.9053, 0.0000, 0.0000, 0.0323, 0.9757,
        0.0000, 1.9665, 1.0276, 0.0000, 0.0588, 0.5085, 0.1956, 0.0000, 0.0000,
        0.0128, 0.0000, 0.0000, 0.0101, 0.9477, 1.1218, 0.0000, 0.0000, 0.3422,
        0.6930, 0.0000, 0.0000])
```

The different types of transfer functions are interchangeable in a neural network architecture. They can be used in different stages, such as the input to the hidden layer or the hidden layer to the output layer, to improve the model's accuracy.

Leaky ReLU

In a standard neural network model, a dying gradient problem is common. To avoid this issue, leaky ReLU is applied. Leaky ReLU allows a small and non-zero gradient when the unit is not active.

```
X = Variable(torch.randn(100, 10))
y = Variable(torch.randn(100, 30))

func = nn.LeakyReLU()
output_x = func(x)
output_y = func(y)
print('Output size : ', output_x.size())
print('Output size : ', output_y.size())

Output size :  torch.Size([100, 10])
Output size :  torch.Size([100, 30])

print(x[0])
print(output_x[0])
print(y[0])
print(output_y[0])
```

```
tensor([ 0.3611, -0.3622,  0.5740, -0.3404, -0.1284,  1.4639,  1.3272,
         0.0636,  -1.1366,  1.1084])
tensor([ 3.6107e-01, -3.6216e-03,  5.7399e-01, -3.4043e-03, -1.2843e-03,
         1.4639e+00,  1.3272e+00,  6.3646e-02, -1.1366e-02,  1.1084e+00])
tensor([-0.4000, -0.2603,  0.5494, -1.1904,  1.0810,  0.0770,  0.5700, -1.0860,
         0.6954, -0.3596, -0.7211, -0.5289,  1.8362, -1.4268, -1.1033,  0.0696,
         0.5678,  0.5952,  0.2172,  0.5269,  1.4032, -0.3520, -0.7009,  0.0710,
        -0.2730, -1.4919, -1.3549,  0.1566, -1.0187,  0.0810])
tensor([-0.0040, -0.0026,  0.5494, -0.0119,  1.0810,  0.0770,  0.5700, -0.0109,
         0.6954, -0.0036, -0.0072, -0.0053,  1.8362, -0.0143, -0.0110,  0.0696,
         0.5678,  0.5952,  0.2172,  0.5269,  1.4032, -0.0035, -0.0070,  0.0710,
        -0.0027, -0.0149, -0.0135,  0.1566, -0.0102,  0.0810])
```

Recipe 4-2. Visualizing the Shape of Activation Functions

Problem

How do you visualize the activation functions? The visualization of activation functions is important in correctly building a neural network model.

Solution

The activation functions translate the data from one layer into another layer. The transformed data can be plotted against the actual tensor to visualize the function. You have taken a sample tensor, converted it to a PyTorch variable, applied the function, and stored it as another tensor. You can represent the actual tensor and the transformed tensor using matplotlib.

How It Works

The right choice of activation function will not only provide better accuracy but also helps with extracting meaningful information.

```
Import torch.nn.functional as F
from torch.autograd import Variable
import matplotlib.pyplot as plt
%matplotlib inline

x = torch.linspace(-10, 10, 1500)
x = Variable(x)
x_1 = x.data.numpy()    # tranforming into numpy

y_relu = F.relu(x).data.numpy()
y_sigmoid = torch.sigmoid(x).data.numpy()
y_tanh = torch.tanh(x).data.numpy()
y_softplus = F.softplus(x).data.numpy()
```

In this script, you have an array in the linear space between –10 and +10 plus 1,500 sample points. You convert the vector to a Torch variable and then make a copy as a NumPy variable for plotting the graph. Then, you calculate the activation functions. Figures 4-1 and 4-4 show the activation functions.

```
Plt.figure(figsize=(7, 4))
plt.plot(x_1, y_relu, c='blue', label='ReLU')
plt.ylim((-1, 11))
plt.legend(loc='best')
```

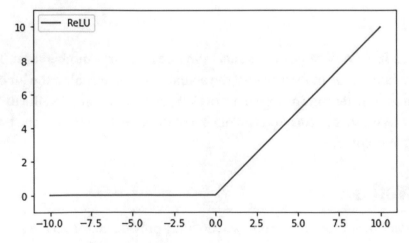

Figure 4-1. *Activation function ReLU*

```
plt.figure(figsize=(7, 4))
plt.plot(x_1, y_sigmoid, c='blue', label='sigmoid')
plt.ylim((-0.2, 1.2))
plt.legend(loc='best')
```

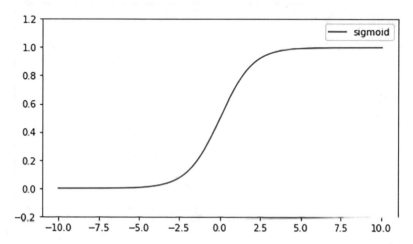

Figure 4-2. *Activation function sigmoid*

```
plt.figure(figsize=(7, 4))
plt.plot(x_1, y_tanh, c='blue', label='tanh')
plt.ylim((-1.2, 1.2))
plt.legend(loc='best')
```

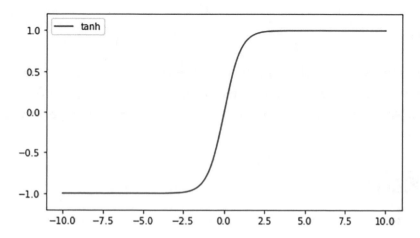

Figure 4-3. *Activation function tanh*

```
plt.figure(figsize=(7, 4))
plt.plot(x_1, y_softplus, c='blue', label='softplus')
plt.ylim((-0.2, 11))
plt.legend(loc='best')
```

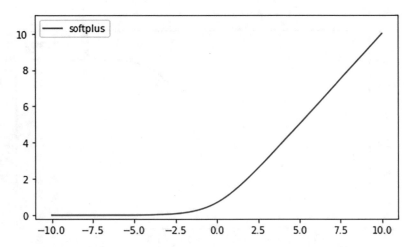

Figure 4-4. *Activation function softplus*

Recipe 4-3. Basic Neural Network Model

Problem

How do you build a basic neural network model using PyTorch?

Solution

A basic neural network model in PyTorch requires six steps: preparing training data, creating a basic neural network model, initializing weights, calculating the loss function, selecting the learning rate, and optimizing the loss function with respect to the model's parameters.

How It Works

Let's follow a step-by-step approach to create a basic neural network model.

```
Def prep_data():
    train_X = np.asarray([13.3,14.4,15.5,16.71,16.93,14.168,19.779,16.182,
                          17.59,12.167,17.042,10.791,15.313,17.997,15.654,
                          19.27,13.1])
    train_Y = np.asarray([11.7,12.76,12.09,13.19,11.694,11.573,13.366,
    12.596, 12.53,11.221,12.827,13.465,11.65,12.904,12.42,12.94,11.3])
    dtype = torch.FloatTensor
    X = Variable(torch.from_numpy(train_X).type(dtype),
                 requires_grad=False).view(17,1)
    y = Variable(torch.from_numpy(train_Y).type(dtype),requires_grad=False)
    return X,y
```

To show a sample neural network model, you prepare the dataset and change the data type to a float tensor. When you work on a project, data preparation for building it is a separate activity. Data preparation should be done in the proper way. In the preceding step, train x and train y are two NumPy vectors. Next, you change the data type to a float tensor because it is necessary for matrix multiplication. The next step is to convert it to variable because a variable has three properties that help you fine-tune the object. In the dataset, you have 17 data points on one dimension.

```
# get dynamic parameters
def set_weights():
    w = Variable(torch.randn(1),requires_grad = True)
    b = Variable(torch.randn(1),requires_grad=True)
    return w,b

#deploy neural network model
def build_network(x):
    y_pred = torch.matmul(x,w)+b
    return y_pred

#implement in PyTorch
import torch.nn as nn
f = nn.Linear(17,1) # Much simpler.
F
Linear(in_features=17, out_features=1, bias=True)
```

The set_weight() function initializes the random weights that the neural network model will use in forward propagation. You need two tensors, weights, and biases. The build_network() function simply multiplies the weights with input, adds the bias to it, and generates the predicted values. This is a custom function that you built. If you need to implement the same thing in PyTorch, it is much simpler to use nn.Linear() when you need to use it for linear regression.

```python
#calculate the loss function
def loss_calc(y,y_pred):
    loss = (y_pred-y).pow(2).sum()
    for param in [w,b]:
        if not param.grad is None: param.grad.data.zero_()
    loss.backward()
    return loss.data

# optimizing results
def optimize(learning_rate):
    w.data -= learning_rate * w.grad.data
    b.data -= learning_rate * b.grad.data
learning_rate = 1e-4
x,y = prep_data()  # x - training data,y - target variables
w,b = set_weights() # w,b - parameters
for i in range(5000):
    y_pred = build_network(x) # function which computes wx + b
    loss = loss_calc(y,y_pred) # error calculation
    if i % 1000 == 0:
        print(loss)
    optimize(learning_rate)    # minimize the loss w.r.t. w, b

tensor(5954.0488)
tensor(44.9320)
tensor(39.5382)
tensor(34.9094)
tensor(30.9371)
```

Once you define a network structure, then you need to compare the results with the output to assess the prediction step. The metric that tracks the accuracy of the system is the loss function, which you want to be minimal. The loss function may have a different shape. How do you know exactly where the loss is at a minimum, which corresponds to which iteration is providing the best results? To know this, you need to apply the optimization function on the loss function; it finds the minimum loss value. Then you can extract the parameters corresponding to that iteration. See Figure 4-5.

```
import matplotlib.pyplot as plt
%matplotlib inline
x_numpy = x.data.numpy()
y_numpy = y.data.numpy()
y_pred = y_pred.data.numpy()
plt.plot(x_numpy,y_numpy,'o')
plt.plot(x_numpy,y_pred,'-')
```

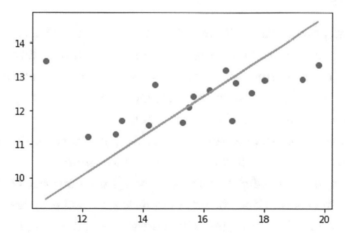

Figure 4-5. *Actual vs. Predicted Tensor*

Standard deviation shows the deviation from the measures of central tendency, which indicates the consistency of the data/variable. It shows whether there is enough fluctuation in data or not.

Recipe 4-4. Tensor Differentiation
Problem

What is tensor differentiation and how is it relevant in computational graph execution using the PyTorch framework?

Solution

The computational graph network is represented by nodes and connected through functions. There are two different kinds of nodes: dependent and independent. *Dependent nodes* wait for results from other nodes to process the input. *Independent nodes* are connected and are either constants or the results. Tensor differentiation is an efficient method to perform computation in a computational graph environment.

How It Works

In a computational graph, tensor differentiation is very effective because the tensors can be computed as parallel nodes, multiprocess nodes, or multithreading nodes. The major deep learning and neural computation frameworks include this tensor differentiation.

Autograd is the function that helps perform tensor differentiation, which means calculating the gradients or slope of the error function and then backpropagating errors through the neural network to fine-tune the weights and biases. Through the learning rate and iteration, it tries to reduce the error value or loss function.

To apply tensor differentiation, the nn.backward() method needs to be applied. Let's take an example and see how the error gradients are backpropagated. To update the curve of the loss function, or to find where the shape of the loss function is minimum and in which direction it is moving, a derivative calculation is required. Tensor differentiation is a way to compute the slope of the function in a computational graph.

```
x = Variable(torch.ones(4, 4) * 12.5, requires_grad=True)
x
tensor([[12.5000, 12.5000, 12.5000, 12.5000], [12.5000, 12.5000, 12.5000,
12.5000], [12.5000, 12.5000, 12.5000, 12.5000], [12.5000, 12.5000, 12.5000,
12.5000]], requires_grad=True)
```

```
fn = 2 * (x * x) + 5 * x + 6

# 2x^2 + 5x + 6
fn.backward(torch.ones(4,4))
print(x.grad)
tensor([[55., 55., 55., 55.],
        [55., 55., 55., 55.],
        [55., 55., 55., 55.],
        [55., 55., 55., 55.]])
```

In this script, the x is a sample tensor for which automatic gradient calculation needs to happen. The fn is a linear function that is created using the x variable. Using the backward function, you can perform a backpropagation calculation. The .grad() function holds the final output from the tensor differentiation.

Conclusion

This chapter discussed various activation functions and the use of the activation functions in various situations. The method or system to select the best activation function is accuracy driven; the activation function that gives the best results should always be used dynamically in the model. You also created a basic neural network model using small sample tensors, updated the weights using optimization, and generated predictions. In the next chapter, you will see more examples.

CHAPTER 5

Supervised Learning Using PyTorch

Supervised machine learning is the most sophisticated branch of machine learning. It is in use in almost all fields, including artificial intelligence, cognitive computing, and language processing. Machine learning literature broadly talks about three types of learning: supervised, unsupervised, and reinforcement learning. In supervised learning, the machine learns to recognize the output; hence, it is task driven and the task can be classification or regression. In unsupervised learning, the machine learns patterns from data; thus, it generalizes new datasets and learning occurs by evaluating a set of input features. In reinforcement learning, the learning happens in response to a system that reacts to situations.

This chapter covers regression techniques in detail with a machine learning approach and interprets the output from regression methods in the context of a business scenario. The algorithmic classification is shown in Figure 5-1.

P. Mishra, *PyTorch Recipes*, https://doi.org/10.1007/978-1-4842-8925-9_5

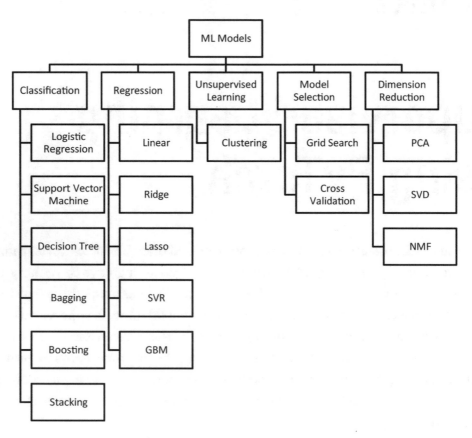

Figure 5-1. *Algorithmic classification*

Each object or row represents one event, and each event is categorized into groups. Identifying which level a record belongs to is called *classification*, in which the target variable has specific labels or tags attached to the events. For example, in a bank database, each customer is tagged as either a loyal customer or not a loyal customer. In a medical records database, each patient's disease is tagged. In the telecom industry, each subscriber is tagged as a churn or non-churn customer. These are examples in which a supervised algorithm performs classification. The word *classification* comes from the classes available in the target column.

In *regression learning*, the objective is to predict the value of a continuous variable. For example, given the features of a property such as the number of bedrooms, square feet, nearby areas, the township, and so forth, the asking price for the house is determined. In such scenarios, regression models can be used. Similar examples include predicting stock prices or the sales, revenue, and profit of a business.

In an unsupervised learning algorithm, there is no outcome variable and tagging or labeling is not available. You are interested in knowing the natural grouping of the observations, or records, or rows in a dataset. This natural grouping should be in such a way that within groups similarity should be at a maximum and between groups similarity should be at a minimum.

In real-world scenarios, there are cases where regression does not help predict the target variable. In supervised regression techniques, the input data is also known as *training data*. For each record, there is a label that has a continuous numerical value. The model is prepared through a training process that predicts the right output, and the process continues until the desired level of accuracy is achieved. You may need advanced regression methods to understand the pattern existing in the dataset.

Introduction to Linear Regression

Linear regression analysis is known as the most reliable, easiest to apply, and most widely used among all statistical techniques. This assumes linear, additive relationships between dependent and independent variables. The objective of linear regression is to predict the dependent or target variable through independent variables. The specification of the linear regression model is as follows:

$$Y = \alpha + \beta X$$

This formula has a property in which the prediction for Y is a straight-line function of each of the X variables, keeping all others fixed, and the contributions of different X variables for the predictions are additive. The slopes of their individual straight-line relationships with Y are the coefficients of the variables. The coefficients and intercept are estimated by least squares (i.e., setting them equal to the unique values that minimize the sum of squared errors within the sample of data to which the model is fitted).

The model's prediction errors are typically assumed to be independently and identically normally distributed. When the beta coefficient becomes zero, the input variable X has no impact on the dependent variable. The ordinary least square (OLS) method attempts to minimize the sum of the squared residuals. The residuals are defined as the difference between the points on the regression line to the actual data points in the scatterplot. This process seeks to estimate the beta coefficients in a multiple linear regression model.

Let's take a sample dataset of 15 people. You capture the height and weight for each of them. By taking only their heights, can you predict the weight of a person using a linear regression technique? The answer is yes.

Person	1	2	3	4	5	6	7	8	9	10	11	12	13	14	15
Height	58	59	60	61	62	63	64	65	66	67	68	69	70	71	72
Weight	115	117	120	123	126	129	132	135	139	142	146	150	154	159	164

To represent this graphically, you measure height on the x axis and you measure weight on the y axis. The linear regression equation is on the graph where the intercept is 87.517 and the coefficient is 3.45. The data points are represented by dots and the connecting line shows linear relationship (see Figure 5-2).

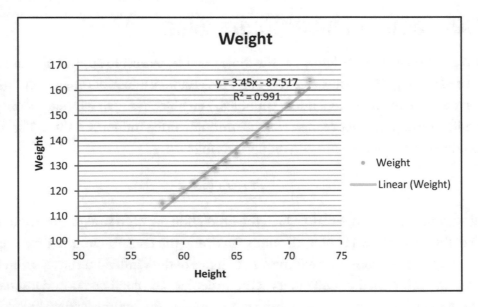

Figure 5-2. *Height and weight relationships*

Why do we assume that a linear relationship exists between the dependent variable and a set of independent variables, when most real-life scenarios reflect any other type of relationship than a linear relationship? The reasons why we stick to linear relationships are described next.

A linear relationship is easy to understand and interpret. There are ways to transform an existing deviation from linearity and make it linear. It is easy to generate predictions.

The field of predictive modeling is mainly concerned with minimizing errors in a predictive model or making the most accurate predictions possible. Linear regression was developed in the field of statistics. It is studied as a model for understanding the relationship between the input and the output of numerical variables, but it has been borrowed by machine learning. It is both a statistical algorithm and a machine learning algorithm. The linear regression model depends on the following set of assumptions:

- The linear relationship between dependent and independent variables.

- There should not be any multicollinearity among the predictors. If you have more than two predictors in the input feature space, the input features should not be correlated.

- There should not be any autocorrelation.

- There should not be any heteroscedasticity. The variance of the error term should be constant, along the predictors on another axis, which means the error variance should be constant.

- The error term should be normally distributed. The error term is basically defined as the difference between an actual and a predicted variable.

Within linear regression, there are different variants, but in machine learning we consider them as one method. For example, if we are using one explanatory variable to predict the dependent variable, it is called a *simple linear regression model.* If we are using more than one explanatory variable, then the model is called a *multiple linear regression model.* The ordinary least square is a statistical technique to predict the linear regression model; hence, sometimes the linear regression model is also known as an *ordinary least square model.*

Linear regression is very sensitive to missing values and outliers because the statistical method of computing a linear regression depends on the mean, standard deviation, and covariance between the variables. The mean is sensitive to outlier values; therefore, it is expected that we need to clear out the outliers before proceeding toward forming the linear regression model.

In machine learning literature, the method for getting optimum beta coefficients that minimize the error in a regression model is achieved by a method called a *gradient descent algorithm*. How does the gradient descent algorithm work? It starts with an initial value, preferably from zero, and updates the scaling factor by a learning rate regularly and iteratively to minimize the error term.

Understanding linear regression based on a machine learning approach requires special data preparation that avoids assumptions by keeping the original data intact. Data transformation is required to make your model more robust.

Recipe 5-1. Data Preparation for a Supervised Model

Problem

How do you perform data preparation for creating a supervised learning model using PyTorch?

Solution

Let's take an open source dataset named mtcars.csv, which is a regression dataset, to test how to create an input and output tensor.

How It Works

First, the necessary library needs to be imported.

```
import torch
import pandas as pd
import numpy as np
import matplotlib.pyplot as plt
from torch.autograd import Variable
import torch.nn.functional as F
%matplotlib inline
torch.__version__
1.12.1+cu113
```

```
df = pd.read_csv("https://raw.githubusercontent.com/pradmishra1/
PublicDatasets/main/mtcars.csv")
del df['Unnamed: 0']
df.head()
```

Model	MPG	Cyl	Disp	HP	Drat	Wt	Qsec	Vs	Am	Gear	Carb
0	Mazda RX4	21.0	6	160.0	110	3.90	2.620	16.46	0	1	4	4
1	Mazda RX4 Wag	21.0	6	160.0	110	3.90	2.875	17.02	0	1	4	4
2	Datsun 710	22.8	4	108.0	93	3.85	2.320	18.61	1	1	4	1
3	Hornet 4 Drive	21.4	6	258.0	110	3.08	3.215	19.44	1	0	3	1
4	Hornet Sportabout	18.7	8	360.0	175	3.15	3.440	17.02	0	0	3	2

The predictor for the supervised algorithm is qsec, which is used to predict the mileage per gallon provided by the car. What is important here is the data type. First, you import the data, which is in NumPy format, into a PyTorch tensor format. The default tensor format is a float. Using the tensor float format would cause errors when performing the optimization function, so it is important to change the tensor data type. You can reformat the tensor type by using the unsqueeze function and specifying that the dimension is equal to 1.

```
torch.manual_seed(1234)    # reproducible
x = torch.unsqueeze(torch.from_numpy(np.array(df.qsec)),dim=1)
y = torch.unsqueeze(torch.from_numpy(np.array(df.mpg)),dim=1)

x[0:10]
tensor([[16.4600], [17.0200], [18.6100], [19.4400], [17.0200], [20.2200],
[15.8400], [20.0000], [22.9000], [18.3000]], dtype=torch.float64)

y[0:10]
tensor([[21.0000], [21.0000], [22.8000], [21.4000], [18.7000], [18.1000],
[14.3000], [24.4000], [22.8000], [19.2000]], dtype=torch.float64)
```

To reproduce the same result, a manual seed needs to be set, so torch.manual_seed(1234) is used. Although you see that the data type is a tensor, if you check the type function, it will show as double because a tensor type double is required for the optimization function.

Recipe 5-2. Forward and Backward PropagationNeural network

Problem

How do you build a neural network torch class function so that you can build a forward propagation method?

Solution

Design the neural network class function, including the hidden layer from the input layer and from the hidden layer to the output layer. In the neural network architecture, the number of neurons in the hidden layer also needs to be specified.

How It Works

In the class Net() function, you first initialize the feature, hidden, and output layers. Then you introduce the back-propagation function using the rectified linear unit as the activation function in the hidden layer.

```
class Net(torch.nn.Module):
    def __init__(self, n_feature, n_hidden, n_output):
        super(Net, self).__init__()
#hidden layer 1 of the neural network
        self.hidden = torch.nn.Linear(n_feature, n_hidden)
#output layer
        self.predict = torch.nn.Linear(n_hidden, n_output)

    def forward(self, x):
        x = F.relu(self.hidden(x)) # activation function for hidden layer
        x = self.predict(x)                # linear output
        return x
```

Figure 5-3 shows the ReLU activation function. It is popularly used across different neural network models; however, the choice of activation function should be based on accuracy. If the accuracy is improved when using with a different activation function, for instance a sigmoid function, you should consider using that.

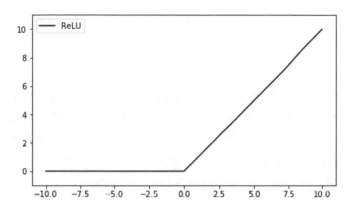

Figure 5-3. *ReLU activation function*

Now the network architecture is mentioned in the supervised learning model. The n_feature shows the number of neurons in the input layer. Since you have one input variable, qsec, you will use 1. The number of neurons in the hidden layer can be decided based on the input and the degree of accuracy required in the learning model. You use the n_hidden equal to 20, which means 20 neurons in the hidden layer 1, and the output neuron is 1.

```
net = Net(n_feature=1, n_hidden=20, n_output=1)
net.double()
print(net)  # Neural network architecture
Net(
  (hidden): Linear(in_features=1, out_features=20, bias=True)
  (predict): Linear(in_features=20, out_features=1, bias=True)
)

optimizer = torch.optim.SGD(net.parameters(), lr=0.2)
loss_func = torch.nn.MSELoss()
# this is for regression mean squared loss
```

The role of the optimization function is to minimize the loss function defined with respect to the parameters and the learning rate. The learning rate chosen here is 0.2. You also pass the neural network parameters into the optimizer. There are various optimization functions:

- SGD: Implements stochastic gradient descent (optionally with momentum). The parameters could be momentum, learning rate, and weight decay.

- Adadelta: Adaptive learning rate. Has five different arguments: parameters of the network, a coefficient used for computing a running average of the squared gradients, the addition of a term for achieving numerical stability of the model, the learning rate, and a weight decay parameter to apply regularization.

- Adagrad: Adaptive subgradient methods for online learning and stochastic optimization. Has arguments such as iterable of parameter to optimize the learning rate and learning rate decay with weight decay.

- Adam: A method for stochastic optimization. This function has six different arguments, an iterable of parameters to optimize, learning rate, betas (known as coefficients used for computing running averages of the gradient and its square), a parameter to improve numerical stability, and so forth.

- ASGD: Acceleration of stochastic approximation by averaging. It has five different arguments, iterable of parameters to optimize, learning rate, decay term, weight decay, and so forth.

- RMSprop algorithm: Uses a magnitude of gradients that are calculated to normalize the gradients.

- SparseAdam: Implements a lazy version of the Adam algorithm suitable for sparse tensors. In this variant, only moments that show up in the gradient are updated, and only those portions of the gradient are applied to the parameters.

Apart from the optimization function, a loss function needs to be selected before running the supervised learning model. Again, there are various loss functions; let's look at the error functions.

- MSELoss: Creates a criterion that measures the mean squared error between elements in the input variable and target variable. For regression-related problems, this is the best loss function.

```
optimizer
SGD ( Parameter Group 0 dampening: 0 foreach: None lr: 0.2 maximize: False
momentum: 0 nesterov: False weight_decay: 0 )
loss_func
MSELoss()
#Turn the interactive mode on
plt.ion()
```

After running the supervised learning model, which is a regression model, you need to print the actual vs. predicted values and represent them in a graphical format; therefore, you need to turn on the interactive feature of the model.

Recipe 5-3. Optimization and Gradient Computation

Problem

How do you build a basic supervised neural network training model using PyTorch with different iterations?

Solution

The basic neural network model in PyTorch requires six different steps: preparing training data, initializing weights, creating a basic network model, calculating loss function, selecting the learning rate, and optimizing the loss function with respect to the parameters of the model.

How It Works

Let's follow a step-by-step approach to create a basic neural network model.

```
for t in range(100):
    prediction = net(x)      # input x and predict based on x
    loss = loss_func(prediction, y)      # must be (1. nn output, 2. target)
    optimizer.zero_grad()    # clear gradients for next train
    loss.backward()          # backpropagation, compute gradients
    optimizer.step()         # apply gradients
```

```
if t % 50 == 0:
    # plot and show learning process
    plt.cla()
    plt.scatter(x.data.numpy(), y.data.numpy())
    plt.plot(x.data.numpy(), prediction.data.numpy(), 'g-', lw=3)
    plt.text(0.5, 0, 'Loss=%.4f' % loss.data.numpy())
    plt.show()
plt.ioff()
```

The final prediction result from the model with the first iteration and the last iteration is now represented in Figure 5-4. In the initial step, the loss function is 276.91. After optimization, the loss function is 35.1890. The fitted regression line and the way it is fitted to the dataset are represented.

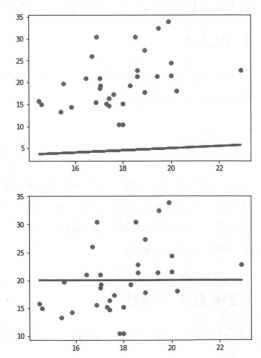

Loss=276.9186

Loss=35.1890

Figure 5-4. *Loss function results*

146

Recipe 5-4. Viewing Predictions
Problem

How do you extract the best results from the PyTorch-based supervised learning model?

Solution

The computational graph network is represented by nodes and connected through functions. Various techniques can be applied to minimize the error function and get the best predictive model. You can increase the iteration numbers, estimate the loss function, optimize the function, print actual and predicted values, and show it all in a graph.

How It Works

To apply tensor differentiation, the nn.backward() method needs to be applied. Let's take an example to see how the error gradients are backpropagated. The grad() function holds the final output from the tensor differentiation. See Figure 5-5.

```
optimizer = torch.optim.SGD(net.parameters(), lr=0.001)
loss_func = torch.nn.MSELoss()  # this is for regression mean squared loss

for t in range(1000):
    prediction = net(x)     # input x and predict based on x
    loss = loss_func(prediction, y) # must be (1. nn output, 2. target)
    optimizer.zero_grad()   # clear gradients for next train
    loss.backward()         # backpropagation, compute gradients
    optimizer.step()        # apply gradients

    if t % 100 == 0:
        # plot and show learning process
        plt.cla()
        plt.scatter(x.data.numpy(), y.data.numpy())
        plt.plot(x.data.numpy(), prediction.data.numpy(), 'g-', lw=3)
        plt.text(0.5, 0, 'Loss=%.4f' % loss.data.numpy())
        plt.show()
plt.ioff() #Turn the interactive mode off
```

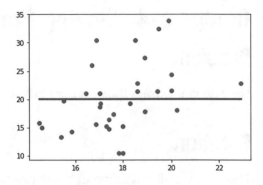

Loss=35.1890

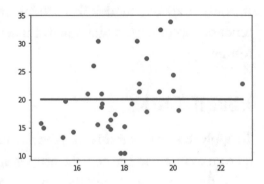

Loss=35.1890

Figure 5-5. *Loss plots*

The tuning parameters that can increase the accuracy of the supervised learning model, which is a regression use case, can be achieved with the following methods:

- Number of iterations

- Type of loss function

- Selection of optimization method

- Selection of loss function

- Learning rate

- Decay in the learning rate

- Momentum require for optimization

The real dataset looks like the following:

```
df.head()
```

148

Model	MPG	Cyl	Disp	HP	Drat	Wt	Qsec	Vs	Am	Gear	Carb
0	Mazda RX4	21.0	6	160.0	110	3.90	2.620	16.46	0	1	4	4
1	Mazda RX4 Wag	21.0	6	160.0	110	3.90	2.875	17.02	0	1	4	4
2	Datsun 710	22.8	4	108.0	93	3.85	2.320	18.61	1	1	4	1
3	Hornet 4 Drive	21.4	6	258.0	110	3.08	3.215	19.44	1	0	3	1
4	Hornet Sportabout	18.7	8	360.0	175	3.15	3.440	17.02	0	0	3	2

The following script explains reading the mpg and qsec columns from the mtcars.csv dataset. It converts those two variables to tensors using the unsqueeze function and then uses it inside the neural network model for prediction.

```
x = torch.unsqueeze(torch.from_numpy(np.array(df.mpg)),dim=1)
y = torch.unsqueeze(torch.from_numpy(np.array(df.qsec)),dim=1)

optimizer = torch.optim.SGD(net.parameters(), lr=0.2)
loss_func = torch.nn.MSELoss()  # this is for regression mean squared loss

plt.ion() #Turn the interactive mode on

for t in range(1000):
    prediction = net(x)     # input x and predict based on x
    loss = loss_func(prediction, y) # must be (1. nn output, 2. target)
    optimizer.zero_grad()   # clear gradients for next train
    loss.backward()         # backpropagation, compute gradients
    optimizer.step()        # apply gradients

    if t % 200 == 0:
        # plot and show learning process
        plt.cla()
        plt.scatter(x.data.numpy(), y.data.numpy())
        plt.plot(x.data.numpy(), prediction.data.numpy(), 'g-', lw=3)
        plt.text(0.5, 0, 'Loss=%.4f' % loss.data.numpy())
        plt.show()
plt.ioff() #Turn the interactive mode off
```

After 1000 iterations, the model converges. See Figure 5-6.

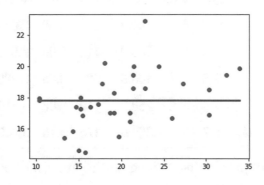

Loss=3.0934

Figure 5-6. *After 1000 iterations, the model converges*

The neural networks in the torch library are typically used with the nn module. Let's take a look at that.

Neural networks can be constructed using the `torch.nn` package, which provides almost all neural network related functionalities, including the following:

- **Linear layers**: `nn.Linear, nn.Bilinear`

- **Convolution layers**: `nn.Conv1d, nn.Conv2d, nn.Conv3d, nn.ConvTranspose2d`

- **Nonlinearities**: `nn.Sigmoid, nn.Tanh, nn.ReLU, nn.LeakyReLU`

- **Pooling layers**: `nn.MaxPool1d, nn.AveragePool2d`

- **Recurrent networks**: `nn.LSTM, nn.GRU`

- **Normalization**: `nn.BatchNorm2d`

- **Dropout**: `nn.Dropout, nn.Dropout2d`

- **Embedding**: `nn.Embedding`

- **Loss functions**: `nn.MSELoss, nn.CrossEntropyLoss, nn.NLLLoss`

The standard classification algorithm is another version of a supervised learning algorithm, in which the target column is a class variable and the features could be numeric and categorical.

Recipe 5-5. Supervised Model Logistic Regression
Problem

How do you deploy a logistic regression model using PyTorch?

Solution

The computational graph network is represented by nodes and connected through functions. Various techniques can be applied to minimize the error function and get the best predictive model. You can increase the iteration numbers, estimate the loss function, optimize the function, print actual and predicted values, and show it all in a graph.

How It Works

To apply tensor differentiation, the `nn.backward()` method needs to be applied. Let's look at an example.

```
import torch
from torch.autograd import Variable
import torch.nn as nn
import torch.nn.functional as F
import matplotlib.pyplot as plt
import torch.optim as optim

torch.manual_seed(1)
```

The following shows data preparation for a logistic regression model:

```
# data preparation for logistic regression
n_data = torch.ones(100,2)
x0 = torch.normal(2*n_data,1)
y0 = torch.zeros(100)

x1 = torch.normal(-2*n_data,1)
y1 = torch.ones(100)
```

```
x = torch.cat((x0,x1),0).type(torch.FloatTensor)
y = torch.cat((y0,y1), ).type(torch.LongTensor)

# Variable conversion
x, y = Variable(x), Variable(y)
# sample data prep for logistic regression model
```

Let's look at the sample dataset for classification. See Figure 5-7.

```
plt.scatter(x.data.numpy()[:,0], x.data.numpy()[:,1],c=y.data.
numpy(),s=100)
plt.show()
```

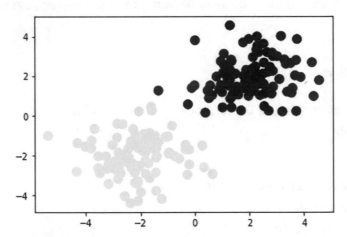

Figure 5-7. *Sample data for logistic regression*

Set up the neural network module for the logistic regression model.

```
class Net(torch.nn.Module):
    def __init__(self, n_feature, n_hidden, n_output):
        super(Net, self).__init__()
# hidden layer
        self.hidden = torch.nn.Linear(n_feature, n_hidden)
#output layer
        self.out = torch.nn.Linear(n_hidden, n_output)
```

```
    def forward(self, x):
#activation function for the hidden layer
        x = F.sigmoid(self.hidden(x))
        x = self.out(x)              # linear output
        return x
```

Check the neural network configuration.

```
net = Net(n_feature=2,n_hidden=10,n_output=2)
print(net)
Net(
  (hidden): Linear(in_features=2, out_features=10, bias=True)
  (out): Linear(in_features=10, out_features=2, bias=True)
)
# loss and optimizer
# softmax is internally computed
# set parameters to be updated
```

Run iterations and find the best solution for the sample graph.

```
#net(x)
optimizer = torch.optim.SGD(net.parameters(),lr=0.02)
loss_func = torch.nn.CrossEntropyLoss()
plt.ion() # interactive graph on

for t in range(100):
    out = net(x)      # input x and predict based on x
    loss = loss_func(out, y)     # must be (1. nn output, 2. target)
    optimizer.zero_grad()   # clear gradients for next train
    loss.backward()         # backpropagation, compute gradients
    optimizer.step()        # apply gradients

    if t % 10 == 0 or t in [3,6]:
        # plot and show learning process
        plt.cla()
        _,prediction = torch.max(F.softmax(out,dim=1),1)
        pred_y = prediction.data.numpy().squeeze()
        target_y = y.data.numpy()
```

```
        plt.scatter(x.data.numpy()[:,0],
                    x.data.numpy()[:,1],
                    c = pred_y,s=100,lw=0)
        accuracy = sum(pred_y == target_y)/200.0

        plt.text(1.5, -4, 'Accuracy=%.2f' % accuracy)
        plt.show()
plt.ioff() #Turn the interactive mode off
```

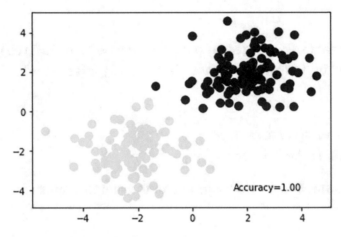

Figure 5-8.   *Scatterplot of actual vs. predicted*

The first iteration provides almost 99% accuracy, and subsequently the model provides 100% accuracy on the training data. See Figures 5-8 and 5-9.

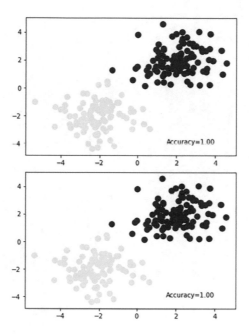

Figure 5-9. *Scatterplots for accuracy*

Final accuracy shows 100, which is a clear case of overfitting, but you can control this by introducing the dropout rate, which was covered in the previous chapter.

Conclusion

This chapter discussed two major types of supervised learning algorithms—linear regression and logistic regression—and their implementation using sample datasets and the PyTorch program. Both algorithms are linear models, one for predicting real valued output and the other for separating one class from another class. Although you considered a two-class classification in the logistic regression example, it can be extended to a multiclass classification model.

CHAPTER 6

Fine-Tuning Deep Learning Models Using PyTorch

Deep learning models have very deep roots in the way biological neurons are connected and the way they transmit information from one node to another node in a network model.

Deep learning has a very specific usage, particularly when single function–based machine learning techniques fail to approximate real-life challenges. For example, when a data dimension is very large (in the thousands), standard machine learning algorithms usually fail to predict or classify the outcome variable. They are also not very efficient computationally. They consume a lot of resources, and model convergence never happens. Most prominent examples are object detection, image classification, and image segmentation.

The most commonly used deep learning algorithms can be classified into three groups.

- **Convolutional neural network**: Mostly suitable for highly sparse datasets, image classification, image recognition, object detection, and so forth.

- **Recurrent neural network**: Applicable to processing sequential information, if there is any internal sequential structure in the way data is generated. This includes music, natural language, audio, and video, where the information is consumed in a sequence.

- **Deep neural network**: Typically applicable when a single layer of a machine learning algorithm cannot classify or predict correctly. There are three variants.

© Pradeepta Mishra 2023
P. Mishra, *PyTorch Recipes*, https://doi.org/10.1007/978-1-4842-8925-9_6

- **Deep network**, where the number of neurons present in each hidden layer is usually more than the previous layer

- **Wide network**, where the number of hidden layers are more than a usual neural network model

- **Both deep and wide network,** where the number of neurons and the number of layers in the network are very high

This chapter discusses how to fine-tune deep learning models using hyperparameters. There is a difference between parameters and hyperparameters. Usually in deep learning models, we are not interested in estimating the parameters because they are the weights and they keep changing based on the initial values, learning rate, and number of iterations. What is important is deciding on the hyperparameters to fine-tune the models, as discussed in Chapter 3, so that optimum results can be derived.

Recipe 6-1. Building Sequential Neural Networks

Problem

Is there any way to build sequential neural network models, as you do in Keras in PyTorch, instead of declaring the neural network models?

Solution

If you declare the entire neural network model, line by line, with the number of neurons, number of hidden layers and iterations, choice of loss functions, optimization functions, selection of weight distribution, and so forth, it will be extremely cumbersome to scale the model. And it is not foolproof—errors could crop up in the model. To avoid issues in declaring the entire model line by line, you can use a high-level function that assumes certain default parameters in the back end and returns the result to the user with minimum hyperparameters. Yes, it is possible to not have to declare the neural network model.

How It Works

Let's look at how to create such models. In the Torch library, the neural network module contains a functional API (application programming interface) that contains various activation functions, as discussed in earlier chapters.

```
import torch
import torch.nn.functional as F
```

In the following lines of script, you create a simple neural network model with a linear function as the activation function for input to the hidden layer and the hidden layer to the output layer.

The following function requires declaring class Net, features, hidden neurons, and activation functions, which can be easily replaced by the sequential module:

```
# replace following class code with an easy sequential network
class Net(torch.nn.Module):
    def __init__(self, n_feature, n_hidden, n_output):
        super(Net, self).__init__()
#hidden layer
        self.hidden = torch.nn.Linear(n_feature, n_hidden)
#output layer
        self.predict = torch.nn.Linear(n_hidden, n_output)

    def forward(self, x):
        x = F.relu(self.hidden(x))# activation function for hidden layer
        x = self.predict(x)             # linear output
        return x
```

Instead of using this script, you can change the class function and replace it with the sequential function. The Keras functions replace the TensorFlow functions, which means that many lines of TensorFlow code can be replaced by a few lines of Keras script. The same thing is possible in PyTorch without requiring any external modules. As an example, in the following, net2 explains the sequential model and net1 explains the preceding script. From a readability perspective, net2 is much better than net1.

```
net1 = Net(1, 100, 1)
# easy and fast way to build your network
net2 = torch.nn.Sequential(
```

```
    torch.nn.Linear(1, 100),
    torch.nn.ReLU(),
    torch.nn.Linear(100, 1)
)
```

If you print both the net1 and net2 model architectures, it does the same thing.

```
print(net1)      # net1 architecture
print(net2)      # net2 architecture

Net(
  (hidden): Linear(in_features=1, out_features=100, bias=True)
  (predict): Linear(in_features=100, out_features=1, bias=True)
)
Sequential(
  (0): Linear(in_features=1, out_features=100, bias=True)
  (1): ReLU()
  (2): Linear(in_features=100, out_features=1, bias=True)
)
```

Recipe 6-2. Deciding the Batch Size
Problem

How do you perform batch data training for a deep learning model using PyTorch?

Solution

Training a deep learning model requires a large amount of labeled data. Typically, it is the process of finding a set of weights and biases in such a way that the loss function becomes minimal with respect to matching the target label. If the training process approximates well to the function, the prediction or classification becomes robust.

How It Works

There are two methods for training a deep learning network: batch training and online training. The choice of training algorithm dictates the method of learning. If the algorithm is backpropagation, then online learning is better. For a deep and wide network model with various layers of backpropagation and forward propagation, then batch training is better.

```
import torch
import torch.utils.data as Data

torch.manual_seed(1234)    # reproducible
```

In the training process, the batch size is 5; you can change the batch size to 8 and see the results. In the online training process, the weights and biases are updated for every training example based on the variations between predicted result and actual result. However, in the batch training process, the differences between actual and predicted values get accumulated and computed as a single number over the batch size and reported at the final layer.

```
BATCH_SIZE = 5

x = torch.linspace(1, 10, 10)      # this is x data (torch tensor)
y = torch.linspace(10, 1, 10)      # this is y data (torch tensor)

torch_dataset = Data.TensorDataset(x, y)
loader = Data.DataLoader(
    dataset=torch_dataset,      # torch TensorDataset format
    batch_size=BATCH_SIZE,      # mini batch size
    shuffle=True,               # random shuffle for training
    num_workers=2,              # subprocesses for loading data
)
```

After training the dataset for five iterations, you can print the batch and step. If you compare online training and batch training, batch training has many more advantages than online training. When the requirement is to train a huge dataset, there are memory constraints. When you cannot process a huge dataset in a CPU environment, batch training comes to the rescue. In a CPU environment, you can process large amounts of data with a smaller batch size.

```
for epoch in range(5):   # train entire dataset 5 times
    for step, (batch_x, batch_y) in enumerate(loader):
# for each training step
        # train your data...
        print('Epoch: ', epoch, '| Step: ', step, '| batch x: ',
              batch_x.numpy(), '| batch y: ', batch_y.numpy())
```

Epoch: 0 | Step: 0 | batch x: [3. 2. 4. 7. 8.] | batch y:
[8. 9. 7. 4. 3.]
Epoch: 0 | Step: 1 | batch x: [10. 6. 5. 9. 1.] | batch y:
[1. 5. 6. 2. 10.]
Epoch: 1 | Step: 0 | batch x: [4. 1. 10. 6. 3.] | batch y:
[7. 10. 1. 5. 8.]
Epoch: 1 | Step: 1 | batch x: [7. 5. 8. 9. 2.] | batch y:
[4. 6. 3. 2. 9.]
Epoch: 2 | Step: 0 | batch x: [6. 1. 2. 5. 9.] | batch y:
[5. 10. 9. 6. 2.]
Epoch: 2 | Step: 1 | batch x: [7. 4. 10. 3. 8.] | batch y:
[4. 7. 1. 8. 3.]
Epoch: 3 | Step: 0 | batch x: [2. 3. 1. 10. 7.] | batch y:
[9. 8. 10. 1. 4.]
Epoch: 3 | Step: 1 | batch x: [9. 6. 8. 4. 5.] | batch y:
[2. 5. 3. 7. 6.]
Epoch: 4 | Step: 0 | batch x: [7. 4. 8. 2. 9.] | batch y:
[4. 7. 3. 9. 2.]
Epoch: 4 | Step: 1 | batch x: [1. 10. 5. 3. 6.] | batch y:
[10. 1. 6. 8. 5.]

Make the batch size as 8 and retrain the model.

```
BATCH_SIZE = 8
loader = Data.DataLoader(
    dataset=torch_dataset,       # torch TensorDataset format
    batch_size=BATCH_SIZE,       # mini batch size
    shuffle=True,                # random shuffle for training
    num_workers=2,               # subprocesses for loading data
)
```

```
for epoch in range(5):   # train entire dataset 5 times
#for each training step
    for step, (batch_x, batch_y) in enumerate(loader):
        # train your data...
        print('Epoch: ', epoch, '| Step: ', step, '| batch x: ',
              batch_x.numpy(), '| batch y: ', batch_y.numpy())
```

Epoch: 0 | Step: 0 | batch x: [7. 2. 5. 8. 1. 4. 6. 3.] | batch y:
[4. 9. 6. 3. 10. 7. 5. 8.]
Epoch: 0 | Step: 1 | batch x: [10. 9.] | batch y: [1. 2.]
Epoch: 1 | Step: 0 | batch x: [5. 1. 7. 8. 10. 9. 6. 3.] | batch
y: [6. 10. 4. 3. 1. 2. 5. 8.]
Epoch: 1 | Step: 1 | batch x: [2. 4.] | batch y: [9. 7.]
Epoch: 2 | Step: 0 | batch x: [6. 2. 3. 1. 8. 7. 5. 10.] | batch
y: [5. 9. 8. 10. 3. 4. 6. 1.]
Epoch: 2 | Step: 1 | batch x: [9. 4.] | batch y: [2. 7.]
Epoch: 3 | Step: 0 | batch x: [4. 3. 5. 7. 2. 10. 6. 1.] | batch
y: [7. 8. 6. 4. 9. 1. 5. 10.]
Epoch: 3 | Step: 1 | batch x: [8. 9.] | batch y: [3. 2.]
Epoch: 4 | Step: 0 | batch x: [5. 7. 8. 10. 3. 2. 4. 9.] | batch
y: [6. 4. 3. 1. 8. 9. 7. 2.]
Epoch: 4 | Step: 1 | batch x: [6. 1.] | batch y: [5. 10.]

Recipe 6-3. Deciding the Learning Rate

Problem

How do you identify the best solution based on the learning rate and number of epochs?

Solution

You take a sample tensor and apply various alternative models and print model parameters. The learning rate and epoch number are associated with model accuracy. To reach the global minimum state of the loss function, it is important to keep the learning rate to a minimum and the epoch number to a maximum so that the iteration can take the loss function to the minimum state.

How It Works

First, the necessary library needs to be imported. To find the minimum loss function, gradient descent is typically used as the optimization algorithm, which is an iterative process. The objective is to find the rate of decline of the loss function with respect to the trainable parameters.

```python
import torch
import torch.utils.data as Data
import torch.nn.functional as F
from torch.autograd import Variable
import matplotlib.pyplot as plt
%matplotlib inline

torch.manual_seed(12345)    # reproducible


LR = 0.01
BATCH_SIZE = 32
EPOCH = 12
```

The sample dataset taken for the experiment includes the following. See Figure 6-1.

```python
# sample dataset
x = torch.unsqueeze(torch.linspace(-1, 1, 1000), dim=1)
y = x.pow(2) + 0.3*torch.normal(torch.zeros(*x.size()))

# plot dataset
plt.scatter(x.numpy(), y.numpy())
plt.show()
```

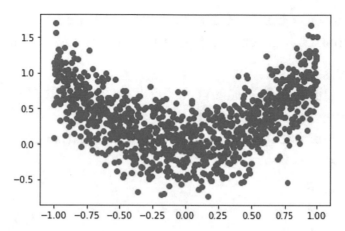

Figure 6-1. *Plotted dataset*

The sample dataset and the first five records look like the following:

```
x[0:10]
tensor([[-1.0000], [-0.9980], [-0.9960], [-0.9940], [-0.9920], [-0.9900],
[-0.9880], [-0.9860], [-0.9840], [-0.9820]])
```

```
y[0:10]
tensor([[0.5561], [1.1422], [0.0882], [1.1212], [1.0920], [0.9764],
[1.0417], [0.5877], [1.6916], [1.5640]])
```

Using the PyTorch utility function, let's load the tensor dataset, introduce the batch size, and test it.

```
torch_dataset = Data.TensorDataset(x, y)
loader = Data.DataLoader(
    dataset=torch_dataset,
    batch_size=BATCH_SIZE,
    shuffle=True, num_workers=2,)
torch_dataset
loader
```

Declare the neural network module.

```
class Net(torch.nn.Module):
    def __init__(self):
        super(Net, self).__init__()
```

```
        self.hidden = torch.nn.Linear(1, 20)    # hidden layer
        self.predict = torch.nn.Linear(20, 1)   # output layer

    def forward(self, x):
        x = F.relu(self.hidden(x))
# activation function for hidden layer
        x = self.predict(x)              # linear output
        return x

net_SGD         = Net()
net_Momentum    = Net()
net_RMSprop     = Net()
net_Adam        = Net()
nets = [net_SGD, net_Momentum, net_RMSprop, net_Adam]

net_Adam
Net( (hidden): Linear(in_features=1, out_features=20, bias=True) (predict):
Linear(in_features=20, out_features=1, bias=True) )

net_Momentum
Net( (hidden): Linear(in_features=1, out_features=20, bias=True) (predict):
Linear(in_features=20, out_features=1, bias=True) )
```

Now, let's look at the network architecture.

```
net_RMSprop
Net( (hidden): Linear(in_features=1, out_features=20, bias=True) (predict):
Linear(in_features=20, out_features=1, bias=True) )

net_SGD
Net( (hidden): Linear(in_features=1, out_features=20, bias=True) (predict):
Linear(in_features=20, out_features=1, bias=True) )
```

While performing the optimization, you can include many options; select the best among the best.

```
opt_SGD         = torch.optim.SGD(net_SGD.parameters(), lr=LR)
opt_Momentum    = torch.optim.SGD(net_Momentum.parameters(),
                            lr=LR, momentum=0.8)
opt_RMSprop     = torch.optim.RMSprop(net_RMSprop.parameters(),
                            lr=LR, alpha=0.9)
```

```
opt_Adam          = torch.optim.Adam(net_Adam.parameters(),
                              lr=LR, betas=(0.9, 0.99))
optimizers = [opt_SGD, opt_Momentum, opt_RMSprop, opt_Adam]

opt_Adam
Adam ( Parameter Group 0 amsgrad: False betas: (0.9, 0.99) capturable:
False eps: 1e-08 foreach: None lr: 0.01 maximize: False weight_decay: 0 )

opt_Momentum
SGD ( Parameter Group 0 dampening: 0 foreach: None lr: 0.01 maximize: False
momentum: 0.8 nesterov: False weight_decay: 0 )

opt_RMSprop
RMSprop ( Parameter Group 0 alpha: 0.9 centered: False eps: 1e-08 foreach:
None lr: 0.01 momentum: 0 weight_decay: 0 )

opt_SGD
SGD ( Parameter Group 0 dampening: 0 foreach: None lr: 0.01 maximize: False
momentum: 0 nesterov: False weight_decay: 0 )

loss_func = torch.nn.MSELoss()
losses_his = [[], [], [], []]   # record loss
loss_func
MSELoss()
```

Recipe 6-4. Performing Parallel Training
Problem

How do you perform parallel data training that includes a lot of models using PyTorch?

Solution

Optimizers are really functions that augment the tensor. The process of finding a best model requires parallel training of many models. The choice of learning rate, batch size, and optimization algorithms make models unique and different from other models. The process of selecting the best model requires hyperparameter optimization.

How It Works

First, the right library needs to be imported. The three hyperparameters (learning rate, batch size, and optimization algorithm) make it possible to train multiple models in parallel, and the best model is decided by the accuracy of the test dataset. The following script uses the stochastic gradient descent algorithm, momentum, RMS prop, and Adam as the optimization method:

```
# training
for epoch in range(EPOCH):
    print('Epoch: ', epoch)
    for step, (batch_x, batch_y) in enumerate(loader):
 # for each training step
        b_x = Variable(batch_x)
        b_y = Variable(batch_y)

        for net, opt, l_his in zip(nets, optimizers, losses_his):
            output = net(b_x)                # get output for every net
            loss = loss_func(output, b_y)  # compute loss for every net
            opt.zero_grad()                # clear gradients for next train
            loss.backward()            # backpropagation, compute gradients
            opt.step()                       # apply gradients
            l_his.append(loss.data)      # loss recoder
labels = ['SGD', 'Momentum', 'RMSprop', 'Adam']
for i, l_his in enumerate(losses_his):
    plt.plot(l_his, label=labels[i])
plt.legend(loc='best')
plt.xlabel('Steps')
plt.ylabel('Loss')
plt.ylim((0, 0.2))
plt.show()
```

Let's look at the chart and epochs.

Epoch: 0
Epoch: 1
Epoch: 2
Epoch: 3
Epoch: 4
Epoch: 5
Epoch: 6
Epoch: 7
Epoch: 8
Epoch: 9
Epoch: 10
Epoch: 11

Out of four optimizers showing loss performance over the steps, the RMSProp optimization method resulted in highest accuracy or minimal loss value. See Figure 6-2.

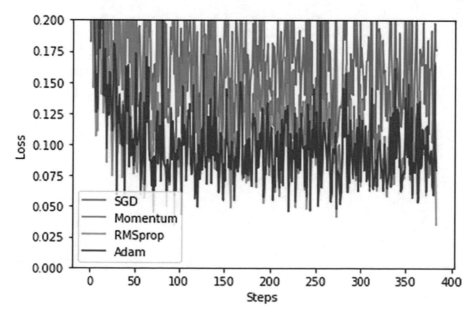

Figure 6-2. *Options*

Conclusion

In this chapter, you looked at various ways to make the deep learning model learn from the training dataset. The training process can be made effective by using hyperparameters. The selection of the right hyperparameter is the key. The deep learning models (convolutional neural network, recurrent neural network, and deep neural network) are different in terms of architecture, but the training process and the hyperparameters remain the same. The choice of hyperparameters and the selection process is much easier in PyTorch than any other framework.

CHAPTER 7

Natural Language Processing Using PyTorch

Natural language processing is the study and evaluation of human language by computers performing various tasks. Natural language study is also known as *computational linguistics*. There are two different components of natural language processing: natural language understanding and natural language generation. *Natural language understanding* involves analysis and knowledge of the input language and responding to it. *Natural language generation* is the process of creating language from input text. Language can be used in various ways. One word may have different meanings, so removing ambiguity is an important part of natural language understanding.

The ambiguity level can be of three types.

- *Lexical ambiguity* is based on parts of speech; deciding whether a word is a noun, verb, adverb, and so forth.

- *Syntactic ambiguity* is where one sentence can have multiple interpretations; the subject and predicate are neutral.

- *Referential ambiguity* is related to an event or scenario expressed in words.

Text analysis is a precursor to natural language processing and understanding. Text analysis means corpus creation (creating a collected set of documents) and then removing white spaces, punctuation, stop words, and junk values such as symbols, emojis, and so forth, which have no textual meaning. After clean-up, the net task is to represent the text in vector form. This is done using the standard Word2vec model, or it can be represented in term frequency and inverse document frequency format (tf-idf). In today's world, we see a lot of applications that use natural language processing; the following are some examples:

© Pradeepta Mishra 2023
P. Mishra, *PyTorch Recipes*, https://doi.org/10.1007/978-1-4842-8925-9_7

- Spell checking applications—online and on smartphones. The user types a particular word, and the system checks the meaning of the word and suggests whether the spelling needs to be corrected.

- Keyword search has been an integral part of our lives over the last decade. Whenever we go to a restaurant, buy something, or visit some place, we do an online search. If the keyword typed is wrong, no match is retrieved; however, the search engine systems are so intelligent that they predict the user's intent and suggest pages that user actually wants to search.

- Predictive text is used in various chat applications. The user types a word, and based on the user's writing pattern, a choice of next words appears. The user is prompted to select any word from the list to frame his sentence.

- Question-and-answering systems like Google Home and Amazon Alexa allow users to interact with the system in natural language. The system processes that information, does an intelligent search, and retrieves the best results for the user.

- Alternate data extraction is when actual data is not available to the user, but the user can use the Internet to fetch data that is publicly available and search for relevant information. For example, if I want to buy a laptop, I want to compare the price of the laptop on various online portals. I have one system scrape the price information from various websites and provide a summary of the prices to me. This process is called *alternate data collection* and it uses web scraping, text processing, and natural language processing.

- Sentiment analysis is a process of analyzing the mood of the customer, user, or agent from the text that they express. It can be used for customer reviews, movie reviews, and so forth. The text presented needs to be analyzed and tagged as a positive sentiment or a negative sentiment. Similar applications can be built using sentiment analysis.

- Topic modeling is the process of finding distinct topics presented in the corpus. For example, we take text from science, math, English, and biology, and jumble all the text, then ask the machine to classify the text and tell us how many topics exist in the corpus, and the machine correctly separates the words present in English from biology, biology from science, and so on so forth. This is called a perfect topic modeling system.

- Text summarization is the process of summarizing the text from the corpus in a shorter format. If we have a two-page document that is 1,000 words, and we need to summarize it in a 200-word paragraph, we can achieve that by using text summarization algorithms.

- Language translation is translating one language to another, such as English to French or French to German. Language translation helps the user understand another language and make the communication process effective.

The study of human language is discrete and very complex. The same sentence may have many meanings, but it is specifically constructed for an intended audience. To understand the complexity of natural language, we not only need tools and programs but also systems and methods. The following five-step approach is used in natural language processing to understand the text from the user.

- Lexical analysis identifies the structure of the word.

- Syntactic analysis is the study of English grammar and syntax.

- Semantic analysis is the meaning of a word in a context.

- PoS (parts of speech) analysis is the understanding and parsing parts of speech.

- Pragmatic analysis is understanding the real meaning of a word in context.

In this chapter, you will use PyTorch to implement the steps that are most commonly used in natural language processing tasks.

Recipe 7-1. Word Embedding
Problem

How do you create a word-embedding model using PyTorch?

Solution

Word embedding is the process of representing words, phrases, and tokens in a meaningful way in a vector structure. The input text is mapped to vectors of real numbers so feature vectors can be used for further computation by machine learning or deep learning models.

How It Works

Words and phrases are represented in real vector format. Words or phrases that have similar meanings in a paragraph or document have similar vector representation. This makes the computation process effective in finding similar words. There are various algorithms for creating embedded vectors from text. Word2vec and GloVe are known frameworks to execute word embeddings. Let's look at the following example:

```
import torch
import torch.nn as nn
import torch.nn.functional as F
import torch.optim as optim

torch.manual_seed(1234)

word_to_ix = {"data": 0, "science": 1}
word_to_ix
{'data': 0, 'science': 1}

embeds = nn.Embedding(2, 5)  # 2 words in vocab, 5 dimensional embeddings
embeds
Embedding(2, 5)

lookup_tensor = torch.tensor([word_to_ix["data"]], dtype=torch.long)
lookup_tensor
tensor([0])
```

The following sets up an embedding layer:

```
hello_embed = embeds(lookup_tensor)
print(hello_embed)
tensor([[ 0.0461,  0.4024, -1.0115,  0.2167, -0.6123]],
       grad_fn=<EmbeddingBackward0>)
```

```
CONTEXT_SIZE = 2
EMBEDDING_DIM = 10
```

Let's look at the sample text. The following text has two paragraphs, and each paragraph has several sentences. If you apply word embedding on these two paragraphs, you will get real vectors as features from the text. These features can be used for further computation.

test_sentence = """The popularity of the term "data science" has exploded in business environments and academia, as indicated by a jump in job openings.[32] However, many critical academics and journalists see no distinction between data science and statistics. Writing in Forbes, Gil Press argues that data science is a buzzword without a clear definition and has simply replaced "business analytics" in contexts such as graduate degree programs.[7] In the question-and-answer section of his keynote address at the Joint Statistical Meetings of American Statistical Association, noted applied statistician Nate Silver said, "I think data-scientist is a sexed up term for a statistician....Statistics is a branch of science. Data scientist is slightly redundant in some way and people shouldn't berate the term statistician."[9] Similarly, in business sector, multiple researchers and analysts state that data scientists alone are far from being sufficient in granting companies a real competitive advantage[33] and consider data scientists as only one of the four greater job families companies require to leverage big data effectively, namely: data analysts, data scientists, big data developers and big data engineers.[34]

On the other hand, responses to criticism are as numerous. In a 2014 Wall Street Journal article, Irving Wladawsky-Berger compares the data science enthusiasm with the dawn of computer science. He argues data science, like any other interdisciplinary field, employs methodologies and practices from across the academia and industry, but then it will morph them into a new discipline. He brings to attention the sharp criticisms computer science, now a well respected academic discipline, had to once face.[35] Likewise, NYU Stern's Vasant Dhar, as do many other academic proponents of data science,[35] argues more specifically in December 2013 that data science is different from the existing practice of data analysis across all disciplines, which focuses only on explaining data sets. Data science seeks actionable and consistent pattern for predictive uses.[1] This practical engineering goal takes data science beyond traditional analytics. Now the data in those disciplines and applied fields that lacked solid theories, like health science and social science, could be sought and utilized to generate powerful predictive models.[1]""".split()

```
# we should tokenize the input, but we will ignore that for now
# build a list of tuples.  Each tuple is ([ word_i-2, word_i-1 ],
target word)
trigrams = [([test_sentence[i], test_sentence[i + 1]], test_
sentence[i + 2])
            for i in range(len(test_sentence) - 2)]
# print the first 3, just so you can see what they look like
print(trigrams[:3])

vocab = set(test_sentence)
word_to_ix = {word: i for i, word in enumerate(vocab)}

[(['The', 'popularity'], 'of'), (['popularity', 'of'], 'the'),
(['of', 'the'], 'term')]
```

Tokenization is the process of splitting sentences into small chunks of tokens, known as *n-grams*. They are called a *unigram* if it is a single word, a *bigram* if it is two words, a *trigram* if it is three words, and so on.

The PyTorch n-gram language modeler can extract relevant key words.

```
class NGramLanguageModeler(nn.Module):

    def __init__(self, vocab_size, embedding_dim, context_size):
        super(NGramLanguageModeler, self).__init__()
        self.embeddings = nn.Embedding(vocab_size, embedding_dim)
        self.linear1 = nn.Linear(context_size * embedding_dim, 128)
        self.linear2 = nn.Linear(128, vocab_size)

    def forward(self, inputs):
        embeds = self.embeddings(inputs).view((1, -1))
```

```
    out = F.relu(self.linear1(embeds))
    out = self.linear2(out)
    log_probs = F.log_softmax(out, dim=1)
    return log_probs

losses = []
loss_function = nn.NLLLoss()
model = NGramLanguageModeler(len(vocab), EMBEDDING_DIM, CONTEXT_SIZE)
optimizer = optim.SGD(model.parameters(), lr=0.001)
```

The n-gram extractor has three arguments: the length of the vocabulary to extract, a dimension of the embedding vector, and context size. Let's look at the loss function and the model specification.

```
model
NGramLanguageModeler( (embeddings): Embedding(228, 10) (linear1):
Linear(in_features=20, out_features=128, bias=Truc) (linear2): Linear(in_
features=128, out_features=228, bias=True) )
```

Apply the Adam optimizer.

```
optimizer
SGD ( Parameter Group 0 dampening: 0 foreach: None lr: 0.001 maximize:
False momentum: 0 nesterov: False weight_decay: 0 )
```

Context extraction from sentences is also important. Let's look at the following function:

```
for epoch in range(10):
    total_loss = 0
    for context, target in trigrams:

        # Step 1. Prepare the inputs to be passed to the model (i.e, turn
          the words
        # into integer indices and wrap them in tensors)
        context_idxs = torch.tensor([word_to_ix[w] for w in context],
        dtype=torch.long)

        # Step 2. Recall that torch *accumulates* gradients. Before
          passing in a
```

```
    # new instance, you need to zero out the gradients from the old
    # instance
    model.zero_grad()

    # Step 3. Run the forward pass, getting log probabilities over next
    # words
    log_probs = model(context_idxs)

    # Step 4. Compute your loss function. (Again, Torch wants
      the target
    # word wrapped in a tensor)
    loss = loss_function(log_probs, torch.tensor([word_to_ix[target]],
    dtype=torch.long))

    # Step 5. Do the backward pass and update the gradient
    loss.backward()
    optimizer.step()

    # Get the Python number from a 1-element Tensor by calling
      tensor.item()
    total_loss += loss.item()
  losses.append(total_loss)
print(losses)  # The loss decreased every iteration over the training data!
```

Recipe 7-2. CBOW Model in PyTorch

Problem

How do you create a CBOW model using PyTorch?

Solution

There are two different methods to represent words and phrases in vectors: *continuous bag of words* (CBOW) and *skip gram*. The bag-of-words approach learns embedding vectors by predicting the word or phrase in context. *Context* means the words before and after the current word. If we take a context of size 4, the four words to the left of the current word and four words to the right of it are considered for context. The model tries to find those eight words in another sentence to predict the current word.

How It Works

Let's look at the following example:

```python
CONTEXT_SIZE = 2  # 2 words to the left, 2 to the right raw_text = """For
the future of data science, Donoho projects an ever-growing environment for
open science where data sets used for academic publications are accessible
to all researchers.[36] US National Institute of Health has already
announced plans to enhance reproducibility and transparency of research
data.[39] Other big journals are likewise following suit.[40][41] This way,
the future of data science not only exceeds the boundary of statistical
theories in scale and methodology, but data science will revolutionize
current academia and research paradigms.[36] As Donoho concludes, "the
scope and impact of data science will continue to expand enormously in
coming decades as scientific data and data about science itself become
ubiquitously available."[36]""".split()

# By deriving a set from `raw_text`, we deduplicate the array
vocab = set(raw_text)
vocab_size = len(vocab)

word_to_ix = {word: i for i, word in enumerate(vocab)}
data = []
for i in range(2, len(raw_text) - 2):
    context = [raw_text[i - 2], raw_text[i - 1],
               raw_text[i + 1], raw_text[i + 2]]
    target = raw_text[i]
    data.append((context, target))
print(data[:5])

class CBOW(nn.Module):

    def __init__(self):
        pass

    def forward(self, inputs):
        pass
# create your model and train.  here are some functions to help you make
```

```
# the data ready for use by your module
def make_context_vector(context, word_to_ix):
    idxs = [word_to_ix[w] for w in context]
    return torch.tensor(idxs, dtype=torch.long)

make_context_vector(data[0][0], word_to_ix)  # example

tensor([26, 54, 63, 18])
```

Graphically, the bag-of-words model looks like Figure 7-1. It has three layers: input, which are the embedding vectors that take the words and phrases into account; the output vector, which is the relevant word predicted by the model; and the projection layer, which is a computational layer provided by the neural network model.

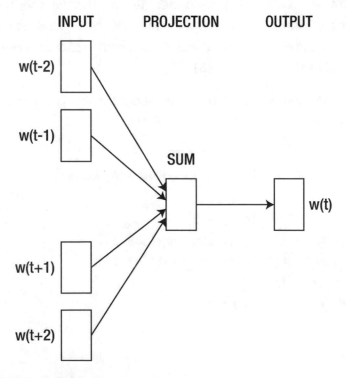

Figure 7-1. *CBOW model representation*

```
lin = nn.Linear(5, 3)  # maps from R^5 to R^3, parameters A, b
# data is 2x5.  A maps from 5 to 3... can we map "data" under A?
data = torch.randn(2, 5)
print(lin(data))  # yes
```

```
data = torch.randn(2, 2)
print(data)
print(F.relu(data))

# Softmax is also in torch.nn.functional
data = torch.randn(5)
print(data)
print(F.softmax(data, dim=0))
print(F.softmax(data, dim=0).sum())   # Sums to 1 because it is a
                                                distribution!
print(F.log_softmax(data, dim=0))  # theres also log_softmax
```

Recipe 7-3. LSTM Model

Problem

How do you create a LSTM model using PyTorch?

Solution

The *long short-term memory* (LSTM) model, also known as the *specific form of recurrent neural network* model, is commonly used in the natural language processing field. Text and sentences come in sequences to make a meaningful sentence, so you need a model that remembers the long and short sequences of text to predict a word or text.

How It Works

Let's look at the following example:

```
lstm = nn.LSTM(3, 3)  # Input dim is 3, output dim is 3
inputs = [torch.randn(1, 3) for _ in range(5)]  # make a sequence of
length 5

# initialize the hidden state.
hidden = (torch.randn(1, 1, 3),
          torch.randn(1, 1, 3))
for i in inputs:
```

```
    # Step through the sequence one element at a time.
    # after each step, hidden contains the hidden state.
    out, hidden = lstm(i.view(1, 1, -1), hidden)
inputs = torch.cat(inputs).view(len(inputs), 1, -1)
hidden = (torch.randn(1, 1, 3), torch.randn(1, 1, 3))  # clean out
                                                         hidden state
out, hidden = lstm(inputs, hidden)
print(out)
print(hidden)

tensor([[[-0.1500,   0.0547,   0.3930]],

        [[-0.1313,  -0.0478,   0.0857]],

        [[-0.1131,   0.0047,  -0.1003]],

        [[ 0.0176,  -0.2464,  -0.1589]],

        [[-0.0523,   0.1781,  -0.1713]]], grad_fn=<StackBackward0>)
(tensor([[[-0.0523,   0.1781,  -0.1713]]], grad_fn=<StackBackward0>),
tensor([[[-0.1997,   0.5137,  -0.6064]]], grad_fn=<StackBackward0>))
```

Prepare a sequence of words as training data to form the LSTM network.

```
def prepare_sequence(seq, to_ix):
    idxs = [to_ix[w] for w in seq]
    return torch.tensor(idxs, dtype=torch.long)

training_data = [
    ("Probability and random variable are integral part of computation
    ".split(),
     ["DET", "NN", "V", "DET", "NN"]),
    ("Understanding of the probability and associated concepts are
    essential".split(),
     ["NN", "V", "DET", "NN"])
]
```

```python
training_data
[(['Probability', 'and', 'random', 'variable', 'are', 'integral', 'part',
'of', 'computation'], ['DET', 'NN', 'V', 'DET', 'NN']), (['Understanding',
'of', 'the', 'probability', 'and', 'associated', 'concepts', 'are',
'essential'], ['NN', 'V', 'DET', 'NN'])]

word_to_ix = {}
for sent, tags in training_data:
    for word in sent:
        if word not in word_to_ix:
            word_to_ix[word] = len(word_to_ix)
print(word_to_ix)
tag_to_ix = {"DET": 0, "NN": 1, "V": 2}

EMBEDDING_DIM = 6
HIDDEN_DIM = 6
class LSTMTagger(nn.Module):

    def __init__(self, embedding_dim, hidden_dim, vocab_size, tagset_size):
        super(LSTMTagger, self).__init__()
        self.hidden_dim = hidden_dim

        self.word_embeddings = nn.Embedding(vocab_size, embedding_dim)

        # The LSTM takes word embeddings as inputs, and
#outputs hidden states with dimensionality hidden_dim.
        self.lstm = nn.LSTM(embedding_dim, hidden_dim)

        # The linear layer that maps from hidden state space to tag space
        self.hidden2tag = nn.Linear(hidden_dim, tagset_size)
        self.hidden = self.init_hidden()

    def init_hidden(self):
        # Before we've done anything, we dont have any hidden state.
        # Refer to the Pytorch documentation to see exactly
        # why they have this dimensionality.
        # The axes semantics are (num_layers, minibatch_size, hidden_dim)
        return (torch.zeros(1, 1, self.hidden_dim),
                torch.zeros(1, 1, self.hidden_dim))
```

```python
    def forward(self, sentence):
        embeds = self.word_embeddings(sentence)
        lstm_out, self.hidden = self.lstm(
            embeds.view(len(sentence), 1, -1), self.hidden)
        tag_space = self.hidden2tag(lstm_out.view(len(sentence), -1))
        tag_scores = F.log_softmax(tag_space, dim=1)
        return tag_scores

model = LSTMTagger(EMBEDDING_DIM, HIDDEN_DIM, len(word_to_ix),
len(tag_to_ix))
loss_function = nn.NLLLoss()
optimizer = optim.SGD(model.parameters(), lr=0.1)
model
loss_function
optimizer
SGD ( Parameter Group 0 dampening: 0 foreach: None lr: 0.1 maximize: False
momentum: 0 nesterov: False weight_decay: 0 )

with torch.no_grad():
    inputs = prepare_sequence(training_data[0][0], word_to_ix)
    tag_scores = model(inputs)
    print(tag_scores)

tensor([[-1.0414, -1.1928, -1.0680],
        [-1.0747, -1.2163, -1.0154],
        [-1.0706, -1.2298, -1.0083],
        [-1.0661, -1.2428, -1.0022],
        [-1.0013, -1.2948, -1.0254],
        [-1.0539, -1.2640, -0.9973],
        [-1.0718, -1.2705, -0.9757],
        [-0.9919, -1.2527, -1.0689],
        [-0.9726, -1.2880, -1.0611]])
```

Summary

This chapter provided recipes on how to apply continuous bag of words, word embedding, and create a long- and short-term memory network. The PyTorch functions corresponding to each recipe can be used for building natural language pipelines for developing solutions such as text classification, automatic text summarization, sentiment analysis, and many other NLP-related processing. The next chapter will cover distributed PyTorch for large scale processing and parallel processing of PyTorch functions and routines. You will learn deep learning model quantization methods for reducing the model size and improving the performance of the model in a deployment use case.

## CHAPTER 8

Distributed PyTorch Modelling, Model Optimization, and Deployment

In this chapter, you will use PyTorch to implement the steps that are most commonly used in installation, training, and setting up distributed PyTorch for model training. The architecture followed for distributed data parallel training and distributed model parallel training can be explained using the following figures. The model optimization process reduces the model parameter's size so that the model object becomes lighter. The bigger the model object, the slower the inference generation. If you reduce the number of layers in the deep learning model, the parameters that are getting trained also lessen, but this may impact the model accuracy. Hence, one technique used to reduce the model size is called *quantization*. There are different types of model quantization that need to be applied in order to put the model into production. Otherwise, bigger model objects are not compatible for deployment.

Recipe 8-1. Distributed Torch Architecture

Problem

What are distributed torch architectures and how are they designed?

Solution

The training load of deep learning models can be spread across multiple GPUs and CPUs as well. There are two ways the spread can happen. One is by distributing the training data and spreading it over multiple processors. Another is by distributing the gradients across multiple processors.

How It Works

In Figure 8-1, the training data samples are distributed by creating smaller batches of data as in mini batch 1 and mini batch 2. A subsample of data is fed to the machines in the cluster, where a number of processors are clubbed together to form a cluster so that the model training can be done in a distributed manner. The deep learning model present in machine 1 has four hidden layers. After mini batch 1's data goes through the four layers, the loss function is estimated. The same process is followed when you feed mini batch 2 to machine 2, having the same exact deep learning model architecture as present in machine 1. Again, the loss estimation happens. Depending upon the loss value, the gradient update goes to the layers in both machines and the updated gradients follow the back propagation method to optimize the model.

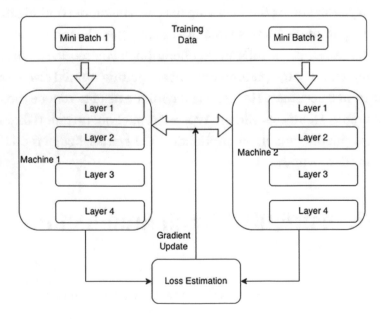

Figure 8-1. *Distributed data parallel training*

In Figure 8-2, the training data is split into mini batches. The mini batch data goes through four hidden layers in machine 1 and then it goes through another set of four hidden layers in machine 2 and then loss estimation happens and the gradient update goes to both machines in parallel. In this process, the gradients are processed in a faster way than the Figure 8-1 method.

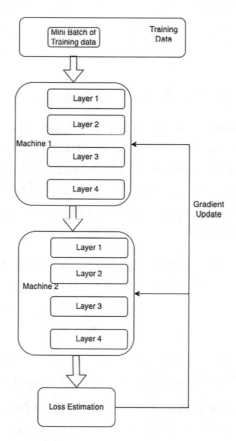

Figure 8-2. *Distributed model parallel training*

Recipe 8-2. Components of Torch Distributed Problem

What are the components of torch distributed?

Solution

The training load of deep learning models can be spread across multiple GPUs and CPUs. There are two ways the spread can happen. One is by distributing the training data and spreading it over multiple processors. Two is by distributing the gradients across multiple processors.

How It Works

There are three main components of a distributed torch framework.

1. **Distributed data parallel training**: This is also known as DDP. In this process, the model architecture is replicated over every process. Every model replica is fed with a different set of input data. In this process, a synchronous method of gradients communication is important and key. This is gracefully managed by the DDP framework

2. **RPC-based distributed training**: Remote procedure call-based distributed training is useful for workloads that cannot be fitted to a DDP framework. An example is parallel distributed pipeline processing.

3. **Collective communication**: There is a library called c10d that provides collective communication and P2P communication.

There are three kinds of backends that can be used: GLOO, NCCL, and MPI. For distributed GPU training, the NCCL backend should be used; for distributed CPU training, the GLOO backend should be used; and if the PyTorch is built from the source, the MPI backend should be used.

Recipe 8-3. Setting Up Distributed PyTorch

Problem

How do you set distributed parallel processing up in PyTorch?

Solution

There are two ways in which you can set up distributed PyTorch: using the GPU on the cloud and using the CPU on a local, single machine. The data parallel training requires a set of processes that need to be followed for optimum utilization of the framework.

How It Works

Let's look at the following example. The import functions add the libraries necessary for installing DDP. The environment setup is done locally on port 55555. The distributed process group requires a backend, which is GLOO in this example. It requires a rank, which should be provided by the user, and the number of worker size, which is world_ size, which also needs to be provided by the user.

```python
import os
import sys
import tempfile
import torch
import torch.distributed as dist
import torch.nn as nn
import torch.optim as optim
import torch.multiprocessing as mp

from torch.nn.parallel import DistributedDataParallel as DDP

def setup(rank, world_size):
    os.environ['MASTER_ADDR'] = 'localhost'
    os.environ['MASTER_PORT'] = '5555'

    # initialize the process group
    dist.init_process_group("gloo", rank=rank, world_size=world_size)

def cleanup():
    dist.destroy_process_group()

class NNET_Model(nn.Module):
    def __init__(self):
        super(NNET_Model, self).__init__()
```

```python
        self.net1 = nn.Linear(10, 10)
        self.relu = nn.ReLU()
        self.net2 = nn.Linear(10, 5)

    def forward(self, x):
        return self.net2(self.relu(self.net1(x)))

def nnet_basic(rank, world_size):
    print(f"Running basic DDP example on rank {rank}.")
    setup(rank, world_size)

    # create model and move it to CPU with id rank
    model = NNET_Model().to(rank)
    ddp_model = DDP(model, device_ids=[rank])

    loss_fn = nn.MSELoss()
    optimizer = optim.SGD(ddp_model.parameters(), lr=0.001)

    optimizer.zero_grad()
    outputs = ddp_model(torch.randn(20, 10))
    labels = torch.randn(20, 5).to(rank)
    loss_fn(outputs, labels).backward()
    optimizer.step()

    cleanup()
```

The basic neural network model has a linear layer that uses DDP in a CPU environment.

```python
nnet_basic(rank=1,world_size=4)
```

This program will take some time because it is running in a CPU environment and world_size=4 means with four workers or processors participating in the training job and rank as 1.

Recipe 8-4. Loading Data to Distributed PyTorch Problem

How do you load datasets into distributed PyTorch?

Solution

The following code shows how to download a MNIST dataset. Using a distributed data sampler, you can prepare the data and train a model.

How It Works

Let's look at the following example:

```
import torch.distributed as dist
def setup(rank, world_size):
    os.environ['MASTER_ADDR'] = 'localhost'
    os.environ['MASTER_PORT'] = '12355'
    dist.init_process_group("nccl", rank=rank, world_size=world_size)

import torchvision.datasets as datasets
mnist_trainset = datasets.MNIST(root='./data', train=True, download=True,
transform=None)

Downloading http://yann.lecun.com/exdb/mnist/train-images-idx3-ubyte.gz
Downloading http://yann.lecun.com/exdb/mnist/train-images-idx3-ubyte.gz to
./data/MNIST/raw/train-images-idx3-ubyte.gz
100%
9912422/9912422 [00:00<00:00, 120503370.11it/s]
Extracting ./data/MNIST/raw/train-images-idx3-ubyte.gz to ./data/MNIST/raw

Downloading http://yann.lecun.com/exdb/mnist/train-labels-idx1-ubyte.gz
Downloading http://yann.lecun.com/exdb/mnist/train-labels-idx1-ubyte.gz to
./data/MNIST/raw/train-labels-idx1-ubyte.gz
100%
28881/28881 [00:00<00:00, 798789.93it/s]
Extracting ./data/MNIST/raw/train-labels-idx1-ubyte.gz to ./data/MNIST/raw

Downloading http://yann.lecun.com/exdb/mnist/t10k-images-idx3-ubyte.gz
Downloading http://yann.lecun.com/exdb/mnist/t10k-images-idx3-ubyte.gz to
./data/MNIST/raw/t10k-images-idx3-ubyte.gz
100%
1648877/1648877 [00:00<00:00, 47304965.85it/s]
```

```
Extracting ./data/MNIST/raw/t10k-images-idx3-ubyte.gz to ./data/MNIST/raw

Downloading http://yann.lecun.com/exdb/mnist/t10k-labels-idx1-ubyte.gz
Downloading http://yann.lecun.com/exdb/mnist/t10k-labels-idx1-ubyte.gz to
./data/MNIST/raw/t10k-labels-idx1-ubyte.gz
100%
4542/4542 [00:00<00:00, 183349.17it/s]
Extracting ./data/MNIST/raw/t10k-labels-idx1-ubyte.gz to ./data/MNIST/raw
```

```python
from torch.utils.data.distributed import DistributedSampler
def prepare(rank, world_size, batch_size=32, pin_memory=False,
num_workers=0):
    dataset = mnist_trainset
    sampler = DistributedSampler(dataset, num_replicas=world_size,
    rank=rank, shuffle=False, drop_last=False)

    dataloader = DataLoader(dataset, batch_size=batch_size, pin_memory=
    pin_memory, num_workers=num_workers, drop_last=False, shuffle=False,
    sampler=sampler)

    return dataloader
```

```python
from torch.nn.parallel import DistributedDataParallel as DDP
def main(rank, world_size):
    # setup the process groups
    setup(rank, world_size)
    # prepare the dataloader
    dataloader = prepare(rank, world_size)

    # instantiate the model(Your Model) and move it to the right device
    model = Model().to(rank)

    # wrap the model with DDP
    # device_ids tell DDP where is your model
    # output_device tells DDP where to output, in our case, it is rank
    # find_unused_parameters=True instructs DDP to find unused output of
        the forward() function of any module in the model
    model = DDP(model, device_ids=[rank], output_device=rank, find_unused_
    parameters=True)
```

```
if torch.cuda.is_available():
    DEVICE = torch.device('cuda')
    device_ids = list(range(torch.cuda.device_count()))
    gpus = len(device_ids)
    print('GPU detected')
else:
    DEVICE = torch.device("cpu")
    print('No GPU. switching to CPU')
```

Recipe 8-5. Quantization of Models in PyTorch

Problem

How do you optimize the deep learning models in PyTorch?

Solution

In order to optimize the deep learning models for efficient deployment on servers and edge devices, PyTorch provides a framework called model quantization. Quantization uses the 8-bit integer format to reduce the weights, which are typically in the form of 32-bit or 64-bit floating points. After applying the quantization, the inference generation from the PyTorch model becomes faster. Quantization can be defined as the technique that is deployed to do computations and memory access with lower precision data.

How It Works

Let's look at the following example. There are three different types of quantization methods available in PyTorch.

- **Dynamic quantization**: There are two set of numerical information that need to be converted into int8: the weights and biases in each hidden layer of a neural network model and the activation functions just before doing computations. Dynamic computation applies quantization at the weights and bias layer as well as on the activations before they get into the computation layer. This is why it is called dynamic; it is applied while model training is going on.

```
import torch.quantization
quantized_model = torch.quantization.quantize_dynamic(model,
                                                {torch.nn.Linear},
                                                dtype=torch.qint8)

print(quantized_model)
ToyModel(
  (net1): DynamicQuantizedLinear(in_features=10, out_features=10,
  dtype=torch.qint8, qscheme=torch.per_tensor_affine)
  (relu): ReLU()
  (net2): DynamicQuantizedLinear(in_features=10, out_features=5,
  dtype=torch.qint8, qscheme=torch.per_tensor_affine)
)
```

- **Static quantization**: This technique is applied after the model object is generated, post training the deep learning model. Static quantization has three major components through which quantization is applied to the model.

  - **Observers**: When the training data is fed into the model, at each activation point the quantization process keeps a check on the resulting distributions of the different activation functions. They are called observers because they observe the change in distribution or statistics around the distribution when quantization is applied.

  - **Fusion**: The operator fusion is the second feature that combines multiple operations into a single operation and thus reduces the volume of computation.

  - **Pre-channel quantization**: This feature helps in quantizing the output channels and speeds up the computation.

    ```
    # insert observers
    torch.quantization.prepare(model, inplace=True)
    # Calibrate the model and collect statistics
    ToyModel( (net1): Linear(in_features=10, out_features=10,
    bias=True) (relu): ReLU() (net2): Linear(in_features=10,
    out_features=5, bias=True) )
    ```

```
# convert to quantized version
torch.quantization.convert(model, inplace=True)
ToyModel( (net1): Linear(in_features=10, out_features=10,
bias=True) (relu): ReLU() (net2): Linear(in_features=10,
out_features=5, bias=True) )
```

- **Quantization-aware training (QAT)**: This method yields higher accuracy because it rounds off the float values to int8 during the forward pass, backward pass, and activation application function; thereby it actually uses the floating point numbers but due to rounding it looks like int8. This results in higher accuracy.

```
# prepare QAT
torch.quantization.prepare_qat(model, inplace=True)

ToyModel( (net1): Linear(in_features=10, out_features=10,
bias=True) (relu): ReLU() (net2): Linear(in_features=10,
out_features=5, bias=True) )

# convert to quantized version, removing dropout, to check
for accuracy on each
epochquantized_model=torch.quantization.convert(model.eval(),
                                          inplace=False)
epochquantized_model
ToyModel( (net1): Linear(in_features=10, out_features=10,
bias=True) (relu): ReLU() (net2): Linear(in_features=10,
out_features=5, bias=True) )
```

Recipe 8-6. Quantization Observer Application Problem

How do you apply different quantization observers in PyTorch?

Solution

There are different types of observers that look at the distributions of activations before computation at each layer. You must understand the behavior of each type of observer and how it works.

How It Works

Let's look at the following example. There are three different types of quantization observer methods available in PyTorch.

- MinMaxObserver

- MovingAverageMinMaxObserver

- HistogramObserver

```
from torch.quantization.observer import MinMaxObserver,
MovingAverageMinMaxObserver, HistogramObserver
C, L = 5, 5
normal = torch.distributions.normal.Normal(0,1)
inputs = [normal.sample((C, L)), normal.sample((C, L))]
print(inputs)
[tensor([[ 0.8052, -0.1585, -1.5735,  0.0400, -0.1424],
        [-1.4450,  1.2916, -0.4354, -1.8434,  0.4686],
        [-1.6375,  0.0545,  0.5203,  0.0024,  0.7699],
        [-1.2877, -2.1810,  0.4022,  1.3470,  0.9177],
        [-0.5629, -0.5823, -1.0329, -1.3076,  0.9457]]),
        tensor([[ 0.3280, -1.9777,  0.2115,  0.8891,  1.2109],
        [-0.0630, -0.4131, -0.3992, -0.4765, -0.7934],
        [ 0.7557, -0.7131, -1.6143, -0.9568,  0.4245],
        [ 0.0509,  0.1589,  0.9872,  1.1071, -0.0961],
        [-0.7442,  1.6635, -0.2982, -0.4168,  0.2499]])]

observers = [MinMaxObserver(), MovingAverageMinMaxObserver(),
HistogramObserver()]
for obs in observers:
  for x in inputs: obs(x)
```

```
print(obs.__class__.__name__, obs.calculate_qparams())
MinMaxObserver (tensor([0.0151]), tensor([145], dtype=torch.int32))
MovingAverageMinMaxObserver (tensor([0.0138]), tensor([157],
dtype=torch.int32))
HistogramObserver (tensor([0.0150]), tensor([143], dtype=torch.int32))

from torch.quantization.observer import
MovingAveragePerChannelMinMaxObserver
obs = MovingAveragePerChannelMinMaxObserver(ch_axis=0)
 # calculate qparams for all `C` channels separately
for x in inputs: obs(x)
print(obs.calculate_qparams())

(tensor([0.0094, 0.0122, 0.0094, 0.0137, 0.0088]), tensor([169, 150, 173,
157, 147], dtype=torch.int32))
```

Recipe 8-7. Quantization Application Using the MNIST Dataset

Problem

How do you apply different quantization techniques on a CNN model using the MNIST dataset in PyTorch?

Solution

As a first step, the dataset need to be loaded to the session. The necessary script for model training need be applied and then the quantization techniques need to be applied.

How It Works

Let's look at the following example:

```
import torch
import torchvision
```

```python
import torchvision.transforms as transforms
import torch.nn as nn
import torch.nn.functional as F
import torch.optim as optim
import os
from torch.utils.data import DataLoader
import torch.quantization
from torch.quantization import QuantStub, DeQuantStub

transform = transforms.Compose(
    [transforms.ToTensor(),
     transforms.Normalize((0.5,), (0.5,))])

trainset = torchvision.datasets.MNIST(root='./data', train=True,
                                      download=True, transform=transform)
trainloader = torch.utils.data.DataLoader(trainset, batch_size=64,
                                          shuffle=True, num_workers=16,
                                          pin_memory=True)

testset = torchvision.datasets.MNIST(root='./data', train=False,
                                     download=True, transform=transform)
testloader = torch.utils.data.DataLoader(testset, batch_size=64,
                                         shuffle=False, num_workers=16,
                                         pin_memory=True)
```

The following class function computes and stores the average and current value of the weights:

```python
class AverageMeter(object):
    """Computes and stores the average and current value"""
    def __init__(self, name, fmt=':f'):
        self.name = name
        self.fmt = fmt
        self.reset()

    def reset(self):
        self.val = 0
        self.avg = 0
```

```
        self.sum = 0
        self.count = 0

    def update(self, val, n=1):
        self.val = val
        self.sum += val * n
        self.count += n
        self.avg = self.sum / self.count

    def __str__(self):
        fmtstr = '{name} {val' + self.fmt + '} ({avg' + self.fmt + '})'
        return fmtstr.format(**self.__dict__)
```

Since you are using the CNN model using the MNIST dataset, you can compute the accuracy using the following function:

```
def accuracy(output, target):
    """ Computes the top 1 accuracy """
    with torch.no_grad():
        batch_size = target.size(0)

        _, pred = output.topk(1, 1, True, True)
        pred = pred.t()
        correct = pred.eq(target.view(1, -1).expand_as(pred))

        res = []
        correct_one = correct[:1].view(-1).float().sum(0, keepdim=True)
        return correct_one.mul_(100.0 / batch_size).item()
```

To print the size of the model object in MB format, you can use the following script:

```
def print_size_of_model(model):
    """ Prints the real size of the model """
    torch.save(model.state_dict(), "temp.p")
    print('Size (MB):', os.path.getsize("temp.p")/1e6)
    os.remove('temp.p')
```

taking the quantized model and real model as two objects together.

```
def load_model(quantized_model, model):
```

```python
    """ Loads in the weights into an object meant for quantization """
    state_dict = model.state_dict()
    model = model.to('cpu')
    quantized_model.load_state_dict(state_dict)

def fuse_modules(model):
    """ Fuse together convolutions/linear layers and ReLU """
    torch.quantization.fuse_modules(model, [['conv1', 'relu1'],
                                            ['conv2', 'relu2'],
                                            ['fc1', 'relu3'],
                                            ['fc2', 'relu4']],
                                            inplace=True)
```

Now you can train the neural network model with convolution layers and three fully connected network layers.

```python
class Net(nn.Module):
    def __init__(self, q = False):
        # By turning on Q we can turn on/off the quantization
        super(Net, self).__init__()
        self.conv1 = nn.Conv2d(1, 6, 5, bias=False)
        self.relu1 = nn.ReLU()
        self.pool1 = nn.MaxPool2d(2, 2)
        self.conv2 = nn.Conv2d(6, 16, 5, bias=False)
        self.relu2 = nn.ReLU()
        self.pool2 = nn.MaxPool2d(2, 2)
        self.fc1 = nn.Linear(256, 120, bias=False)
        self.relu3 = nn.ReLU()
        self.fc2 = nn.Linear(120, 84, bias=False)
        self.relu4 = nn.ReLU()
        self.fc3 = nn.Linear(84, 10, bias=False)
        self.q = q
        if q:
            self.quant = QuantStub()
            self.dequant = DeQuantStub()
```

```python
    def forward(self, x: torch.Tensor) -> torch.Tensor:
        if self.q:
          x = self.quant(x)
        x = self.conv1(x)
        x = self.relu1(x)
        x = self.pool1(x)
        x = self.conv2(x)
        x = self.relu2(x)
        x = self.pool2(x)
        # Be careful to use reshape here instead of view
        x = x.reshape(x.shape[0], -1)
        x = self.fc1(x)
        x = self.relu3(x)
        x = self.fc2(x)
        x = self.relu4(x)
        x = self.fc3(x)
        if self.q:
          x = self.dequant(x)
        return x
```

The trained model object size is 0.178587MB.

```python
net = Net(q=False)
print_size_of_model(net)
Size (MB): 0.178587
```

The average meter that you defined above computes the loss and accuracy in runtime. You run 100 epochs because the model is not running in an GPU environment.

```python
def train(model: nn.Module, dataloader: DataLoader, cuda=False, q=False):
    criterion = nn.CrossEntropyLoss()
    optimizer = optim.SGD(model.parameters(), lr=0.001, momentum=0.9)
    model.train()
    for epoch in range(20):  # loop over the dataset multiple times

        running_loss = AverageMeter('loss')
        acc = AverageMeter('train_acc')
        for i, data in enumerate(dataloader, 0):
```

```
            # get the inputs; data is a list of [inputs, labels]
            inputs, labels = data
            if cuda:
              inputs = inputs.cuda()
              labels = labels.cuda()

            # zero the parameter gradients
            optimizer.zero_grad()

            if epoch>=3 and q:
              model.apply(torch.quantization.disable_observer)

            # forward + backward + optimize
            outputs = model(inputs)
            loss = criterion(outputs, labels)
            loss.backward()
            optimizer.step()

            # print statistics
            running_loss.update(loss.item(), outputs.shape[0])
            acc.update(accuracy(outputs, labels), outputs.shape[0])
            if i % 100 == 0:    # print every 100 mini-batches
                print('[%d, %5d] ' %
                    (epoch + 1, i + 1), running_loss, acc)
    print('Finished Training')
```

The final epoch has run and finished training. As a next step, use the test script.

```
def test(model: nn.Module, dataloader: DataLoader, cuda=False) -> float:
    correct = 0
    total = 0
    model.eval()
    with torch.no_grad():
        for data in dataloader:
            inputs, labels = data

            if cuda:
              inputs = inputs.cuda()
              labels = labels.cuda()
```

```
        outputs = model(inputs)
        _, predicted = torch.max(outputs.data, 1)
        total += labels.size(0)
        correct += (predicted == labels).sum().item()

    return 100 * correct / total
train(net, trainloader)
[15,    301]  loss 0.044805 (0.041544) train_acc 98.437500 (98.681478)
[15,    401]  loss 0.089017 (0.040428) train_acc 98.437500 (98.753117)
[15,    501]  loss 0.001939 (0.041203) train_acc 100.000000 (98.727545)
[15,    601]  loss 0.031541 (0.042560) train_acc 98.437500 (98.679285)
[15,    701]  loss 0.047192 (0.042918) train_acc 96.875000 (98.684914)
[15,    801]  loss 0.011530 (0.043959) train_acc 100.000000 (98.642322)
[15,    901]  loss 0.030178 (0.044269) train_acc 98.437500 (98.638665)
[16,      1]  loss 0.006916 (0.006916) train_acc 100.000000 (100.000000)
score = test(net, testloader, cuda=False)
print('Accuracy of the network on the test images: {}% - FP32'.
format(score))
Accuracy of the network on the test images: 98.65% - FP32
```

The accuracy of the CNN model is 98.65% in floating point weight type. Now, apply the quantization.

```
qnet = Net(q=True)
load_model(qnet, net)
fuse_modules(qnet)
print_size_of_model(qnet)
score = test(qnet, testloader, cuda=False)
print('Accuracy of the fused network on the test images: {}% - FP32'.
format(score))
Size (MB): 0.178843
Accuracy of the fused network on the test images: 98.65% - FP32
```

After applying fuse, the model object remains 0.178843MB and the accuracy remains as it is.

```
qnet.qconfig = torch.quantization.default_qconfig
print(qnet.qconfig)
```

```
torch.quantization.prepare(qnet, inplace=True)
print('Post Training Quantization Prepare: Inserting Observers')
print('\n Conv1: After observer insertion \n\n', qnet.conv1)

test(qnet, trainloader, cuda=False)
print('Post Training Quantization: Calibration done')
torch.quantization.convert(qnet, inplace=True)
print('Post Training Quantization: Convert done')
print('\n Conv1: After fusion and quantization \n\n', qnet.conv1)
print("Size of model after quantization")
print_size_of_model(qnet)
QConfig(activation=functools.partial(<class 'torch.ao.quantization.
observer.MinMaxObserver'>, quant_min=0, quant_max=127){}, weight=functools.
partial(<class 'torch.ao.quantization.observer.MinMaxObserver'>,
dtype=torch.qint8, qscheme=torch.per_tensor_symmetric){})
Post Training Quantization Prepare: Inserting Observers

 Conv1: After observer insertion


 ConvReLU2d(
   (0): Conv2d(1, 6, kernel_size=(5, 5), stride=(1, 1), bias=False)
   (1): ReLU()
   (activation_post_process): MinMaxObserver(min_val=inf, max_val=-inf)
 )
Post Training Quantization: Calibration done
Post Training Quantization: Convert done

 Conv1: After fusion and quantization

 QuantizedConvReLU2d(1, 6, kernel_size=(5, 5), stride=(1, 1),
scale=0.06902680546045303, zero_point=0, bias=False)
Size of model after quantization
Size (MB): 0.049714

score = test(qnet, testloader, cuda=False)
print('Accuracy of the fused and quantized network on the test images:
{}% - INT8'.format(score))
```

```
Accuracy of the fused and quantized network on the test images:
98.58% - INT8
```

The model size is reduced significantly after applying the quantization: the model size becomes 0.05MB without reducing the accuracy of the model. Sometimes quantization can reduce the accuracy. Now let's change the default observers and apply custom observers.

```
from torch.quantization.observer import MovingAverageMinMaxObserver

qnet = Net(q=True)
load_model(qnet, net)
fuse_modules(qnet)

qnet.qconfig = torch.quantization.QConfig(
                        activation=MovingAverageMinMaxObserver.with_
                        args(reduce_range=True),
                        weight=MovingAverageMinMaxObserver.with_args
                        (dtype=torch.qint8, qscheme=torch.per_tensor_symmetric))
print(qnet.qconfig)
torch.quantization.prepare(qnet, inplace=True)
print('Post Training Quantization Prepare: Inserting Observers')
print('\n Conv1: After observer insertion \n\n', qnet.conv1)

test(qnet, trainloader, cuda=False)
print('Post Training Quantization: Calibration done')
torch.quantization.convert(qnet, inplace=True)
print('Post Training Quantization: Convert done')
print('\n Conv1: After fusion and quantization \n\n', qnet.conv1)
print("Size of model after quantization")
print_size_of_model(qnet)
score = test(qnet, testloader, cuda=False)
print('Accuracy of the fused and quantized network on the test images:
{}% - INT8'.format(score))

QConfig(activation=functools.partial(<class 'torch.ao.quantization.
observer.MovingAverageMinMaxObserver'>, reduce_range=True){},
weight=functools.partial(<class 'torch.ao.quantization.observer.
```

```
MovingAverageMinMaxObserver'>, dtype=torch.qint8, qscheme=torch.per_tensor_
symmetric){})
```

Post Training Quantization Prepare: Inserting Observers

 Conv1: After observer insertion

```
ConvReLU2d(
  (0): Conv2d(1, 6, kernel_size=(5, 5), stride=(1, 1), bias=False)
  (1): ReLU()
  (activation_post_process): MovingAverageMinMaxObserver(min_val=inf,
  max_val=-inf)
)
```
```
/usr/local/lib/python3.7/dist-packages/torch/ao/quantization/observer.
py:178: UserWarning: Please use quant_min and quant_max to specify the
range for observers.                        reduce_range will be deprecated in
                                            a future release of PyTorch.
  reduce_range will be deprecated in a future release of PyTorch."
```
Post Training Quantization: Calibration done
Post Training Quantization: Convert done

 Conv1: After fusion and quantization

```
 QuantizedConvReLU2d(1, 6, kernel_size=(5, 5), stride=(1, 1),
scale=0.06884118169546127, zero_point=0, bias=False)
```
Size of model after quantization
Size (MB): 0.049714
Accuracy of the fused and quantized network on the test images:
98.6% - INT8

Now apply QAT (quantization-aware training). With it, all weights and activations are falsely quantized as floats by rounding the numbers. The additional change in the quantization config is required in order to improve the accuracy.

```
qnet = Net(q=True)
load_model(qnet, net)
fuse_modules(qnet)
```

```
qnet.qconfig = torch.quantization.get_default_qconfig('fbgemm')
print(qnet.qconfig)

torch.quantization.prepare(qnet, inplace=True)
test(qnet, trainloader, cuda=False)
torch.quantization.convert(qnet, inplace=True)
print("Size of model after quantization")
print_size_of_model(qnet)
QConfig(activation=functools.partial(<class 'torch.ao.quantization.
observer.HistogramObserver'>, reduce_range=True){}, weight=functools.
partial(<class 'torch.ao.quantization.observer.PerChannelMinMaxObserver'>,
dtype=torch.qint8, qscheme=torch.per_channel_symmetric){})
Size of model after quantization
Size (MB): 0.055572
```

Now, after quantization, the model size is reduced without impacting the accuracy of the model.

```
score = test(qnet, testloader, cuda=False)
print('Accuracy of the fused and quantized network on the test images:
{}% - INT8'.format(score))
Accuracy of the fused and quantized network on the test images:
98.58% - INT8
```

After applying the fusion and quantization both to the convolution layer 1:

```
qnet = Net(q=True)
fuse_modules(qnet)
qnet.qconfig = torch.quantization.get_default_qat_qconfig('fbgemm')
torch.quantization.prepare_qat(qnet, inplace=True)
print('\n Conv1: After fusion and quantization \n\n', qnet.conv1)
qnet=qnet
Conv1: After fusion and quantization

 ConvReLU2d(
  1, 6, kernel_size=(5, 5), stride=(1, 1), bias=False
  (weight_fake_quant): FusedMovingAvgObsFakeQuantize(
```

```
    fake_quant_enabled=tensor([1]), observer_enabled=tensor([1]),
    scale=tensor([1.]), zero_point=tensor([0], dtype=torch.int32),
    dtype=torch.qint8, quant_min=-128, quant_max=127, qscheme=torch.
    per_channel_symmetric, reduce_range=False
    (activation_post_process): MovingAveragePerChannelMinMaxObserver(min_
    val=tensor([]), max_val=tensor([]))
  )
  (activation_post_process): FusedMovingAvgObsFakeQuantize(
    fake_quant_enabled=tensor([1]), observer_enabled=tensor([1]),
    scale=tensor([1.]), zero_point=tensor([0], dtype=torch.int32),
    dtype=torch.quint8, quant_min=0, quant_max=127, qscheme=torch.
    per_tensor_affine, reduce_range=True
    (activation_post_process): MovingAverageMinMaxObserver(min_val=inf,
max_val=-inf)
  )
)

train(qnet, trainloader, cuda=False)


[15,   301]  loss 0.024571 (0.050655) train_acc 100.000000 (98.421927)
[15,   401]  loss 0.083018 (0.052274) train_acc 96.875000 (98.394638)
[15,   501]  loss 0.137679 (0.053624) train_acc 96.875000 (98.334581)
[15,   601]  loss 0.067262 (0.053238) train_acc 98.437500 (98.328307)
[15,   701]  loss 0.025749 (0.053818) train_acc 100.000000 (98.334968)
[15,   801]  loss 0.100811 (0.054027) train_acc 96.875000 (98.341916)
[15,   901]  loss 0.040471 (0.053821) train_acc 98.437500 (98.336917)
[16,     1]  loss 0.024462 (0.024462) train_acc 100.000000 (100.000000)
[16,   101]  loss 0.047599 (0.050869) train_acc 96.875000 (98.452970)

qnet = qnet.cpu()
torch.quantization.convert(qnet, inplace=True)
print("Size of model after quantization")
print_size_of_model(qnet)

score = test(qnet, testloader, cuda=False)
print('Accuracy of the fused and quantized network (trained quantized) on
the test images: {}% - INT8'.format(score))
```

```
Size of model after quantization
Size (MB): 0.055572
Accuracy of the fused and quantized network (trained quantized) on the test
images: 98.69% - INT8
```

The size of the model is reduced to 0.55MB after quantization and fused application.

```
qnet = Net(q=True)
fuse_modules(qnet)
qnet.qconfig = torch.quantization.get_default_qat_qconfig('fbgemm')
torch.quantization.prepare_qat(qnet, inplace=True)
qnet = qnet
train(qnet, trainloader, cuda=False, q=True)
qnet = qnet.cpu()
torch.quantization.convert(qnet, inplace=True)
print("Size of model after quantization")
print_size_of_model(qnet)

score = test(qnet, testloader, cuda=False)
print('Accuracy of the fused and quantized network (trained quantized) on
the test images: {}% - INT8'.format(score))
[15,   601]  loss 0.070189 (0.060619) train_acc 98.437500 (98.128120)
[15,   701]  loss 0.275282 (0.060933) train_acc 93.750000 (98.123217)
[15,   801]  loss 0.029783 (0.060075) train_acc 100.000000 (98.172207)
[15,   901]  loss 0.028618 (0.059222) train_acc 98.437500 (98.186043)
[16,     1]  loss 0.046160 (0.046160) train_acc 98.437500 (98.437500)
[16,   101]  loss 0.110157 (0.069398) train_acc 96.875000 (98.004332)
[16,   201]  loss 0.157249 (0.063327) train_acc 95.312500 (98.173197)
[16,   301]  loss 0.013502 (0.059892) train_acc 100.000000 (98.183140)

Finished Training
Size of model after quantization
Size (MB): 0.055572
Accuracy of the fused and quantized network (trained quantized) on the test
images: 98.53% - INT8
```

It is safe to conclude that after quantization the model size is significantly reduced without reducing the accuracy of the model in this particular example. This is not always going to be true. There must be a step-by-step process to apply quantization and observe the delta change in the accuracy parameter. The reason why you want to apply quantization is to hasten the inference generation time. If the accuracy gets compromised, there is no point in applying quantization. Hence quantization-aware training usually provides accuracy closer to the floating point accuracy.

Summary

In this chapter, you explored recipes for applying distributed PyTorch in a GPU environment and processing the model training in a parallel way. Also, you saw recipes that provide ways to quantize large deep learning model objects into a smaller size, keeping accuracy in mind. Quantization is necessary to hasten the inference generation time for deep learning models. We discussed various methods of quantization in this chapter. The next chapter will cover data augmentation methods for image and audio data and feature engineering and extraction of relevant features using PyTorch.

## CHAPTER 9

Data Augmentation, Feature Engineering, and Extractions for Image and Audio

In an audio classification model, you want the deep learning algorithm to learn sounds and predict a category. Similarly, in image classification, you want the deep learning model to remember the images, learn the patterns from the images, and classify new images into various categories that the learning algorithm has been trained on. In sound classification, you typically start by taking the audio files as inputs and convert them into something called a spectogram. A spectrogram produces a high-dimensional space of data that can be further reduced by applying the convolutional neural network model. As you know, the final layer of a CNN is a neural network, which is also called a fully connected layer and is typically used as a classifier.

In this chapter, you will use PyTorch to implement the steps that are most commonly used in audio processing and image processing tasks.

© Pradeepta Mishra 2023
P. Mishra, *PyTorch Recipes*, https://doi.org/10.1007/978-1-4842-8925-9_9

Recipe 9-1. Spectogram for Audio Processing

Problem

How do you augment the audio data for training using a spectogram?

Solution

The raw audio files can be in mp4, mp3, or wav format, but you cannot use them directly for model training. The model training process requires the data to be in a structured and tabular format. So, the problem here is how do you transform the audio files into a tabular format?

How It Works

The audio data transformation for application of a deep learning model can be separated into several steps. First, the audio files need to be loaded from a wav file into memory (depending upon the environment, CPU, or GPU). Second, the wav file needs to be converted into stereo. Time-shifting audio augmentation happens and then the audio files get converted into a spectrogram. Generally, audio files are of two channels because of stereo but sometimes the audio clips are from one channel only. Hence a normalization process is also required here before training so that the input data can be standardized.

The torchaudio module has two sub-components to process audio data:

- Torchaudio.functional
- Torchaudio.transforms

```
import torch
import torchaudio
import torchaudio.functional as F
import torchaudio.transforms as T

print(torch.__version__)
print(torchaudio.__version__)

from IPython.display import Audio
```

```python
import librosa
import matplotlib.pyplot as plt
from torchaudio.utils import download_asset

torch.random.manual_seed(0)
SAMPLE_SPEECH = download_asset("YOURSAMPLE_AUDIO_DATA.wav")
```

The following function converts the audio file to a display of a wave form:

```python
def plot_waveform(waveform, sr, title="Waveform"):
    waveform = waveform.numpy()

    num_channels, num_frames = waveform.shape
    time_axis = torch.arange(0, num_frames) / sr

    figure, axes = plt.subplots(num_channels, 1)
    axes.plot(time_axis, waveform[0], linewidth=1)
    axes.grid(True)
    figure.suptitle(title)
    plt.show(block=False)
```

This function converts the input data into a spectrogram:

```python
def plot_spectrogram(specgram, title=None, ylabel="freq_bin"):
    fig, axs = plt.subplots(1, 1)
    axs.set_title(title or "Spectrogram (db)")
    axs.set_ylabel(ylabel)
    axs.set_xlabel("frame")
    im = axs.imshow(librosa.power_to_db(specgram), origin="lower",
    aspect="auto")
    fig.colorbar(im, ax=axs)
    plt.show(block=False)
```

The torchaudio functional module is called *stateless* because it can be implemented as a standalone function.

Recipe 9-2. Installation of Torchaudio

Problem

How do you install torchaudio using PyTorch?

Solution

The torchaudio module works with a particular version as a date. It has issues with default versions, so please use a specific version while using PIP for installation. After installation, restart your Google Colab runtime environment or Python Jupyter environment.

How It Works

The audio-related functions from PyTorch work when you install and import the torchaudio module.

```
pip install torchaudio==0.4.0
Looking in indexes: https://pypi.org/simple, https://us-python.pkg.dev/
colab-wheels/public/simple/
Collecting torchaudio==0.4.0
  Downloading torchaudio-0.4.0-cp37-cp37m-manylinux1_x86_64.whl (3.1 MB)
     |████████████████████████████████| 
3.1 MB 8.6 MB/s
Collecting torch==1.4.0
  Downloading torch-1.4.0-cp37-cp37m-manylinux1_x86_64.whl (753.4 MB)
     |████████████████████████████████| 
753.4 MB 6.4 kB/s
Installing collected packages: torch, torchaudio
  Attempting uninstall: torch
    Found existing installation: torch 1.7.0+cpu
    Uninstalling torch-1.7.0+cpu:
      Successfully uninstalled torch-1.7.0+cpu
  Attempting uninstall: torchaudio
    Found existing installation: torchaudio 0.7.0
```

```
    Uninstalling torchaudio-0.7.0:
        Successfully uninstalled torchaudio-0.7.0
ERROR: pip's dependency resolver does not currently take into account
all the packages that are installed. This behaviour is the source of the
following dependency conflicts.
torchvision 0.8.1+cpu requires torch==1.7.0, but you have torch 1.4.0 which
is incompatible.
torchtext 0.13.1 requires torch==1.12.1, but you have torch 1.4.0 which is
incompatible.
fastai 2.7.9 requires torch<1.14,>=1.7, but you have torch 1.4.0 which is
incompatible.
fastai 2.7.9 requires torchvision>=0.8.2, but you have torchvision
0.8.1+cpu which is incompatible.
Successfully installed torch-1.4.0 torchaudio-0.4.0
```

WARNING: The following packages were previously imported in this runtime:
  [torch]
You must restart the runtime in order to use newly installed versions.

Please note the last line as restart is mandatory. If that does not happen, you may get an error.

```
import torchaudio
import torchaudio.functional as F
import torchaudio.transforms as T

print(torch.__version__)
print(torchaudio.__version__)
```

Recipe 9-3. Loading Audio Files into PyTorch
Problem

How do you load data to torchaudio using PyTorch?

Solution

The torchaudio module has built-in datasets that can be used to train deep learning models. However, the real challenge is to load any other file as raw and load it into torchaudio and apply transformations.

How It Works

The following script explains how to load data from a built-in library and from a local directory. See Figure 9-1.

```
yesno_data = torchaudio.datasets.YESNO('.', download=True)
data_loader = torch.utils.data.DataLoader(yesno_data,
                                          batch_size=1,
                                          shuffle=True,
                                          num_workers=2)
data_loader
```

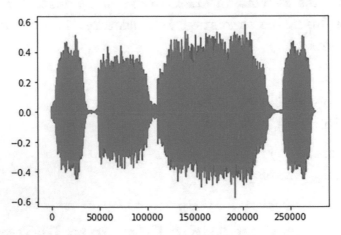

Figure 9-1. *Waveform sample data*

In order to read from an external URL, given audio data, do the following:

```
audio_url = "https://pytorch.org/tutorials/_static/img/steam-train-whistle-
daniel_simon-converted-from-mp3.wav"
```

```
request_url = requests.get(audio_url)

with open('steam-train-whistle-daniel_simon-converted-from-mp3.wav', 'wb')
as file:

    file.write(request_url.content)
audio_file = "steam-train-whistle-daniel_simon-converted-from-mp3.wav"

data_waveform, rate_of_sample = torchaudio.load(audio_file)
print("This is the shape of the waveform: {}".format(data_waveform.size()))

print("This is the output for Sample rate of the waveform: {}".format(rate_
of_sample))
This is the shape of the waveform: torch.Size([2, 276858])
This is the output for Sample rate of the waveform: 44100

plt.figure()

plt.plot(data_waveform.t().numpy())
```

Recipe 9-4. Installation of Librosa for Audio

Problem

How do you install librosa for sound data transformation using PyTorch?

Solution

Feature extraction from audio and sound files and performing transformations requires a set of functions that is provided by librosa. It is a Python package.

How It Works

First, start with a new notebook or Colab notebook.

```
!pip install librosa
Looking in indexes: https://pypi.org/simple, https://us-python.pkg.dev/
colab-wheels/public/simple/
```

```
Requirement already satisfied: librosa in /usr/local/lib/python3.7/dist-
packages (0.8.1)
Requirement already satisfied: decorator>=3.0.0 in /usr/local/lib/
python3.7/dist-packages (from librosa) (4.4.2)
Requirement already satisfied: scipy>=1.0.0 in /usr/local/lib/python3.7/
dist-packages (from librosa) (1.7.3)
Requirement already satisfied: numba>=0.43.0 in /usr/local/lib/python3.7/
dist-packages (from librosa) (0.56.0)
Requirement already satisfied: packaging>=20.0 in /usr/local/lib/python3.7/
dist-packages (from librosa) (21.3)
Requirement already satisfied: audioread>=2.0.0 in /usr/local/lib/
python3.7/dist-packages (from librosa) (3.0.0)
Requirement already satisfied: resampy>=0.2.2 in /usr/local/lib/python3.7/
dist-packages (from librosa) (0.4.0)
Requirement already satisfied: soundfile>=0.10.2 in /usr/local/lib/
python3.7/dist-packages (from librosa) (0.10.3.post1)
Re
from IPython.display import Audio
import librosa
import matplotlib.pyplot as plt
from torchaudio.utils import download_asset

torch.random.manual_seed(0)

SAMPLE_SPEECH = download_asset("/content/waves_yesno/0_0_0_0_1_1_1_1.wav")
```

Just for visualization you can use the following. See Figure 9-2.

```
SPEECH_WAVEFORM, SAMPLE_RATE = torchaudio.load(SAMPLE_SPEECH)

plot_waveform(SPEECH_WAVEFORM, SAMPLE_RATE, title="Original waveform")
Audio(SPEECH_WAVEFORM.numpy(), rate=SAMPLE_RATE)
```

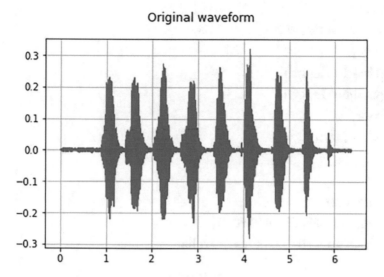

Figure 9-2. *Original waveform*

Recipe 9-5. Spectogram Transformation
Problem

How do you create spectrograms from sound files?

Solution

Spectrograms can be defined as a visual representation of the spectrum of frequencies of a signal with time varying fluctuations. They are also called *voicegrams*.

How It Works

The audio data transformation for the application of a deep learning model can be separated into several processes, including a spectrogram. See Figure 9-3.

```
import torchaudio.transforms as T
n_fft = 1024
win_length = None
hop_length = 512
```

```
# Define transform
spectrogram = T.Spectrogram(
    n_fft=n_fft,
    win_length=win_length,
    hop_length=hop_length,
    center=True,
    pad_mode="reflect",
    power=2.0,
)
# Perform transform
spec = spectrogram(SPEECH_WAVEFORM)
plot_spectrogram(spec[0], title="torchaudio")
```

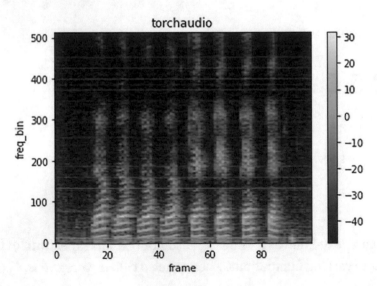

Figure 9-3.   *Torch audio frame*

Recipe 9-6. Griffin-Lim Transformation

Problem

How do you apply a Griffin-Lim transformation?

Solution

The Griffin-Lim algorithm (GLA) helps in making the spectrogram consistent by iterating two projections where a spectrogram is expected to be in its inter-bin. A consistent spectrogram that maintains its amplitude is a need for signals, so GLA transformation is necessary for data augmentation when you want to recover the waveform from a spectogram.

How It Works

The GLA can be applied by using the following script. See Figure 9-4.

```
import torchaudio.transforms as T

torch.random.manual_seed(0)

n_fft = 1024
win_length = None
hop_length = 512

spec = T.Spectrogram(
    n_fft=n_fft,
    win_length=win_length,
    hop_length=hop_length,
)(SPEECH_WAVEFORM)
griffin_lim = T.GriffinLim(
    n_fft=n_fft,
    win_length=win_length,
    hop_length=hop_length,
)
reconstructed_waveform = griffin_lim(spec)
plot_waveform(reconstructed_waveform, SAMPLE_RATE, title="Reconstructed")
Audio(reconstructed_waveform, rate=SAMPLE_RATE)
```

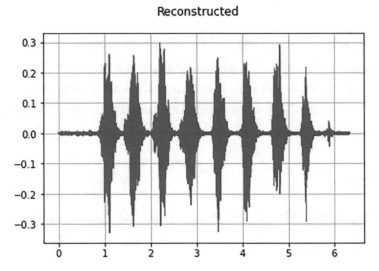

Figure 9-4. *Reconstructed waveform*

Recipe 9-7. Mel Scale Transformation Using a Filter Bank

Problem

How do you apply filter bank for converting the frequency bins to mel scale bins?

Solution

In order to transform frequency bins to mel scale bins, the torchaudio functional module provides a filter bank. This does not require input audio features. A filter bank can be defined as method of discretizing a continuous frequency response into various bins. The type of filterbank depends upon the use case. There are various filters and a mel filter bank is one.

How It Works

You can leverage the mel scale transformation from torchaudio. See Figure 9-5.

```
import torchaudio.transforms as T
n_fft = 255
n_mels = 61
sample_rate = 5000

mel_filters = T.melscale_fbanks(
    int(n_fft // 2 + 1),
    n_mels=n_mels,
    f_min=0.0,
    f_max=sample_rate / 2.0,
    sample_rate=sample_rate,
    norm="slaney",
)
plot_fbank(mel_filters, "Mel Filter Bank - torchaudio")
```

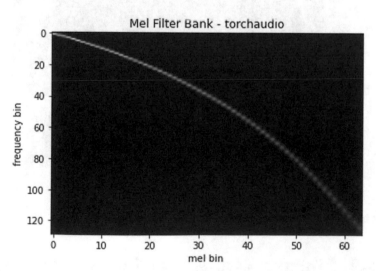

Figure 9-5. *Mel Filter Bank from torch audio*

The same thing can be achieved using the librosa library, which does a similar application of filters as those present in torchaudio. See Figure 9-6.

```
#Librosa mel filter
mel_filters_librosa = librosa.filters.mel(
    sr=sample_rate,
    n_fft=n_fft,
    n_mels=n_mels,
    fmin=0.0,
    fmax=sample_rate / 2.0,
    norm="slaney",
    htk=True,
).T
plot_fbank(mel_filters_librosa, "Mel Filter Bank - librosa")

mse = torch.square(mel_filters - mel_filters_librosa).mean().item()
print("Mean Square Difference: ", mse)
```

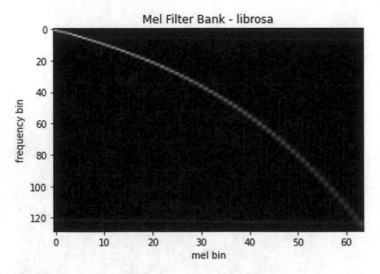

Figure 9-6. *Mel Filter Bank, librosa*

Recipe 9-8. Librosa Mel Scale Conversion vs. the Torchaudio Version

Problem

How do you compare the librosa mel scale conversion to torchaudio using PyTorch?

Solution

The torchaudio module contains transforms that can be developed after generating a spectrum. The mel spectrogram requires the sample rate, window length, hop length padding, and power. Given such parameters from a voice sample as a waveform, a spectrogram can be generated. See Figure 9-7.

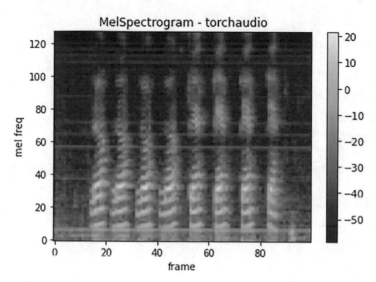

Figure 9-7. *Mel Spectrogram*

How It Works

The audio data transformation can be done using librosa as well as the torchaudio library.

```
import torchaudio.transforms as T
n_fft = 1024
win_length = None
hop_length = 512
n_mels = 128

mel_spectrogram = T.MelSpectrogram(
    sample_rate=sample_rate,
    n_fft=n_fft,
    win_length=win_length,
```

```
        hop_length=hop_length,
        center=True,
        pad_mode="reflect",
        power=2.0,
        norm="slaney",
        onesided=True,
        n_mels=n_mels,
        mel_scale="htk",
    )

melspec = mel_spectrogram(SPEECH_WAVEFORM)
plot_spectrogram(melspec[0], title="MelSpectrogram - torchaudio",
ylabel="mel freq")
```

Now you can generate the mel scale spectrogram with librosa also. See Figure 9-8.

```
melspec_librosa = librosa.feature.melspectrogram(
    y=SPEECH_WAVEFORM.numpy()[0],
    sr=sample_rate,
    n_fft=n_fft,
    hop_length=hop_length,
    win_length=win_length,
    center=True,
    pad_mode="reflect",
    power=2.0,
    n_mels=n_mels,
    norm="slaney",
    htk=True,
)
plot_spectrogram(melspec_librosa, title="MelSpectrogram - librosa",
ylabel="mel freq")

mse = torch.square(melspec - melspec_librosa).mean().item()
print("Mean Square Difference: ", mse)
```

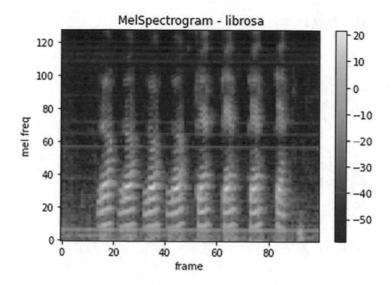

Figure 9-8.

Recipe 9-9. MFCC and LFCC Using Librosa and Torchaudio

Problem

How do we apply Mel Frequency Cepstral Coefficients (MFCC), Linear Frequency Cepstral Coefficients (LFCC) to augment the voice data?

Solution

Torchaudio module contains various data augmentation routines, these methods depends on the kind of algorithm we are going to choose and apply on the voice data. for example if someone wants to apply linear models like Gaussian Mixture Models then first they have to get MFCC from applying Discrete Cosine Transformation (DCT) on the mel-spectogram.

How It Works

MFCC is a compressed representation suitable for linear models with limited data, but if the data size is more and classification use case is there then convolutional neural network works and for that mel-spectogram is better.

```
import torchaudio.transforms as T
n_fft = 2048
win_length = None
hop_length = 512
n_mels = 256
n_mfcc = 256

mfcc_transform = T.MFCC(
    sample_rate=sample_rate,
    n_mfcc=n_mfcc,
    melkwargs={
        "n_fft": n_fft,
        "n_mels": n_mels,
        "hop_length": hop_length,
        "mel_scale": "htk",
    },
)

mfcc = mfcc_transform(SPEECH_WAVEFORM)
plot_spectrogram(mfcc[0])
```

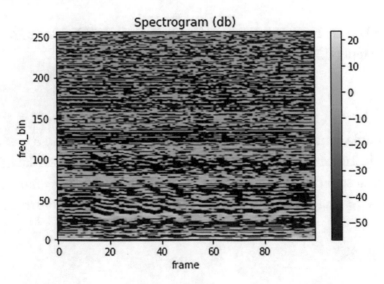

Using Librosa similar spectrogram can be achieved.

```
melspec = librosa.feature.melspectrogram(
    y=SPEECH_WAVEFORM.numpy()[0],
    sr=sample_rate,
    n_fft=n_fft,
    win_length=win_length,
    hop_length=hop_length,
    n_mels=n_mels,
    htk=True,
    norm=None,
)

mfcc_librosa = librosa.feature.mfcc(
    S=librosa.core.spectrum.power_to_db(melspec),
    n_mfcc=n_mfcc,
    dct_type=2,
    norm="ortho",
)
plot_spectrogram(mfcc_librosa)

mse = torch.square(mfcc - mfcc_librosa).mean().item()
print("Mean Square Difference: ", mse)
```

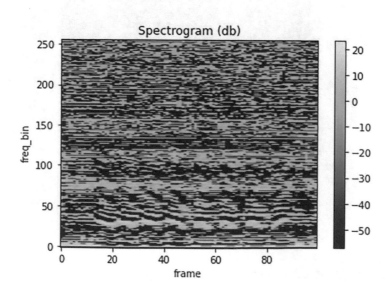

The following code snippet explains the LFCC implementation as a transformation technique.

```
import torchaudio.transforms as T
n_fft = 2048
win_length = None
hop_length = 512
n_lfcc = 256

lfcc_transform = T.LFCC(
    sample_rate=sample_rate,
    n_lfcc=n_lfcc,
    speckwargs={
        "n_fft": n_fft,
        "win_length": win_length,
        "hop_length": hop_length,
    },
)

lfcc = lfcc_transform(SPEECH_WAVEFORM)
plot_spectrogram(lfcc[0])
```

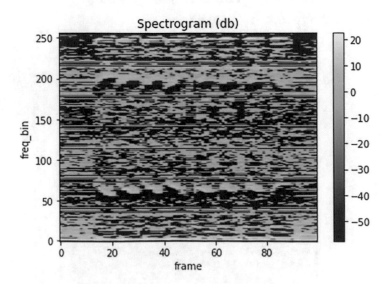

The same expression can be achieved by using librosa as well.

Recipe 9-10. Data Augmentation for Images

Problem

How do you augment image data by applying transforms using PyTorch?

Solution

You are going to use the CIFAR10 dataset to see how to apply the transforms and compose function for data augmentation for images. See Figures 9-9 through 9-11.

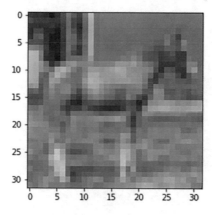

Figure 9-9.

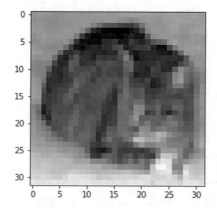

Figure 9-10.

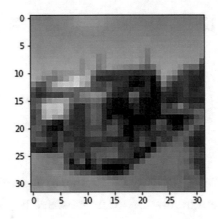

Figure 9-11.

How It Works

The following code snippet shows the application of filters:

```
import os
import sys
import random
import tempfile
import torch
import torch.distributed as dist
import torch.nn as nn
import torch.optim as optim
import torch.multiprocessing as mp
import torchvision
import torchvision.datasets as datasets
import torchvision.transforms as transforms
# Import dataset and dataloaders related packages
from torchvision import datasets
from torchvision.transforms import ToTensor
from torch.utils.data import DataLoader
from torchvision.transforms import Compose, Grayscale

# Download and load the images from the CIFAR10 dataset
cifar10_data = datasets.CIFAR10(
    root="data",  # path where the images will be stored
```

```
    download=True,  # all images should be downloaded
    transform=ToTensor()  # transform the images to tensors
    )

# Print the number of samples in the loaded dataset
print(f"Number of samples: {len(cifar10_data)}")
print(f"Class names: {cifar10_data.classes}")
Files already downloaded and verified
Number of samples: 50000
Class names: ['airplane', 'automobile', 'bird', 'cat', 'deer', 'dog',
'frog', 'horse', 'ship', 'truck']

# Choose a random sample
random.seed(2021)
image, label = cifar10_data[random.randint(0, len(cifar10_data))]
print(f"Label: {cifar10_data.classes[label]}")
print(f"Image size: {image.shape}")
Label: horse
Image size: torch.Size([3, 32, 32])

import matplotlib.pyplot as plt
plt.imshow(image.permute(1, 2, 0))
plt.show()
data = datasets.CIFAR10(root="data", download=True,

                        transform=Compose([ToTensor(), Grayscale()]))
  # Display a random grayscale image
image, label = data[random.randint(0, len(data))]
plt.imshow(image.squeeze(), cmap="gray")
plt.show()
# Load the training samples

training_data = datasets.CIFAR10(
    root="data",
    train=True,
    download=True,
    transform=ToTensor()
    )
```

```python
# Load the test samples
test_data = datasets.CIFAR10(
    root="data",
    train=False,
    download=True,
    transform=ToTensor()
    )
# Create dataloaders with
train_dataloader = DataLoader(training_data, batch_size=64, shuffle=True)
test_dataloader = DataLoader(test_data, batch_size=64, shuffle=True)

# Load the next batch
batch_images, batch_labels = next(iter(train_dataloader))
print('Batch size:', batch_images.shape)

# Display the first image from the batch
plt.imshow(batch_images[0].permute(1, 2, 0))
plt.show()
transform = transforms.Compose(

    [transforms.ToTensor(),
      transforms.Normalize((0.5,), (0.5,))])
```

Following the above normalizing as a data sugmentation technique, random resized crop and random horizontal flip can be applied on the raw images. These additional data augmentation scripts are available on the official documentation page of PyTorch.

Conclusion

In this chapter, you explored data augmentation techniques for audio and images, audio transformation such as waveform transformation, image filtering, and augmentation. In the next chapter, you are going to learn about the libraries scorch and Captum. Scorch provides a routine to apply Scikit-learn functions and APIs such as pipeline, grid search, and cross validation on top of PyTorch models. Captum provides an option to run model interpretability on top of deep learning models developed using the PyTorch framework.

## CHAPTER 10

PyTorch Model Interpretability and Interface to Sklearn

Model interpretability is an area that needs special attention because it is connected with model adoption in particular and AI adoption in general. Users will adopt a model and framework if they can explain the decisions or predictions generated by the deep learning model. In this chapter, you will explore a new framework called Captum, which consists of a set of algorithms that can explain or help us interpret the predictions, model results, and layers of a neural network model. In this chapter, you are also going to use another framework called skorch, which is a library compatible for Scikit-learn users. Machine learning users prefer the sklearn library to train models, perform grid searches, and identify the best hyper parameters of the models—the same kind of seamless experience the users can experience when developing deep neural network models using PyTorch.

Different kinds of interpretability methods are embedded in the Captum library to helps explain deep learning models developed using the PyTorch framework. A neural network interpretation can be done knowing the feature importance, dominance layer identification, and dominant neuron identification. Captum provides three attribution algorithms that help in achieving this information.

- **Primary attribution**: Helps interpret feature importance

- **Layer attribution**: Helps identify the contribution of each neuron in a given layer to the output of the model

- **Neuron attribution**: Helps identify each input feature on the activation of a neuron

© Pradeepta Mishra 2023
P. Mishra, *PyTorch Recipes*, https://doi.org/10.1007/978-1-4842-8925-9_10

In this chapter, you'll use scorch and Captum to implement the steps that are most commonly used in model interpretability and sklearn compatibility.

Recipe 10-1. Installation of Captum
Problem

How do you install Captum?

Solution

There are two ways to install Captum: either using the sudo command or using the pip command.

How It Works

The following syntax explains how to install the library:

```
conda install captum -c pytorch
```

   OR

```
pip install captum
```

```
Looking in indexes: https://pypi.org/simple, https://us-python.pkg.dev/
colab-wheels/public/simple/
Collecting captum
  Downloading captum-0.5.0-py3-none-any.whl (1.4 MB)
     |████████████████████████████████| 1.4 MB 29.2 MB/s
Requirement already satisfied: numpy in /usr/local/lib/python3.7/dist-
packages (from captum) (1.21.6)
Requirement already satisfied: matplotlib in /usr/local/lib/python3.7/dist-
packages (from captum) (3.2.2)
Requirement already satisfied: torch>=1.6 in /usr/local/lib/python3.7/dist-
packages (from captum) (1.12.1+cu113)
Requirement already satisfied: typing-extensions in /usr/local/lib/
python3.7/dist-packages (from torch>=1.6->captum) (4.1.1)
```

```
Requirement already satisfied: python-dateutil>=2.1 in /usr/local/lib/
python3.7/dist-packages (from matplotlib->captum) (2.8.2)
Requirement already satisfied: pyparsing!=2.0.4,!=2.1.2,!=2.1.6,>=2.0.1 in
/usr/local/lib/python3.7/dist-packages (from matplotlib->captum) (3.0.9)
Requirement already satisfied: cycler>=0.10 in /usr/local/lib/python3.7/
dist-packages (from matplotlib->captum) (0.11.0)
Requirement already satisfied: kiwisolver>=1.0.1 in /usr/local/lib/
python3.7/dist-packages (from matplotlib->captum) (1.4.4)
Requirement already satisfied: six>=1.5 in /usr/local/lib/python3.7/dist-
packages (from python-dateutil>=2.1->matplotlib->captum) (1.15.0)
Installing collected packages: captum
Successfully installed captum-0.5.0
```

A safe and easy way to install is via Anaconda. There is a Python version dependency and a PyTorch version dependency. The Python version need to be equal to or over 3.6 and the PyTorch version need to be equal to or over 1.2.

Recipe 10-2. Primary Attribution Feature Importance of a Deep Learning Model
Problem

How do you implement primary attribution using Captum?

Solution

The primary attribution layer provides integrated gradients (IG), gradient shapely additive explanations (SHAP,) saliency, and more to interpret the model better. You are going to use the popular `titanic.csv` dataset to develop a classification model using PyTorch and later on apply primary attribution using IG and other layers. The `titanic` dataset contains features and representations on who did and did not survive the disaster. Using the output column `survived` you will develop a classification model.

How It Works

The following syntax explains how to achieve the attribution layer. IG represents the integral of the gradients with respect to the inputs along the path from the input.

```python
# Initial imports
import numpy as np

import torch

from captum.attr import IntegratedGradients
from captum.attr import LayerConductance
from captum.attr import NeuronConductance

import matplotlib
import matplotlib.pyplot as plt
%matplotlib inline

from scipy import stats
import pandas as pd
dataset_path = "https://raw.githubusercontent.com/pradmishra1/
PublicDatasets/main/titanic.csv"

titanic_data = pd.read_csv(dataset_path)
del titanic_data['Unnamed: 0']
del titanic_data['PassengerId']
titanic_data = pd.concat([titanic_data,
                          pd.get_dummies(titanic_data['Sex']),
                          pd.get_dummies(titanic_data['Embarked'],prefix=
                          "embark"),
                          pd.get_dummies(titanic_data['Pclass'],prefix="cla
                          ss")], axis=1)
titanic_data["Age"] = titanic_data["Age"].fillna(titanic_data["Age"].mean())
titanic_data["Fare"] = titanic_data["Fare"].fillna(titanic_data["Fare"].mean())
titanic_data = titanic_data.drop(['Name','Ticket','Cabin','Sex','Embarked',
'Pclass'], axis=1)

# Set random seed for reproducibility.
np.random.seed(707)
```

```
# Convert features and labels to numpy arrays.
labels = titanic_data["Survived"].to_numpy()
titanic_data = titanic_data.drop(['Survived'], axis=1)
feature_names = list(titanic_data.columns)
data = titanic_data.to_numpy()

# Separate training and test sets using
train_indices = np.random.choice(len(labels), int(0.7*len(labels)),
replace=False)
test_indices = list(set(range(len(labels))) - set(train_indices))
train_features = data[train_indices]
train_labels = labels[train_indices]
test_features = data[test_indices]
test_labels = labels[test_indices]
train_features.shape
(623, 12)
```

There are 623 records that can be used to train the model and 12 features that can be used for model training. The neural network module from torch is used to create 12 hidden neurons in first hidden layer and another 12 neurons in second hidden layer and finally the target variable has two labels as outcome. Two linear hidden layers with a sigmoid activation function are applied. In the final layer, the softmax activation function is applied in order to get the class probabilities.

```
import torch
import torch.nn as nn
torch.manual_seed(1)  # Set seed for reproducibility.
class TitanicSimpleNNModel(nn.Module):
    def __init__(self):
        super().__init__()
        self.linear1 = nn.Linear(12, 12)
        self.sigmoid1 = nn.Sigmoid()
        self.linear2 = nn.Linear(12, 8)
        self.sigmoid2 = nn.Sigmoid()
        self.linear3 = nn.Linear(8, 2)
        self.softmax = nn.Softmax(dim=1)
```

```python
    def forward(self, x):
        lin1_out = self.linear1(x)
        sigmoid_out1 = self.sigmoid1(lin1_out)
        sigmoid_out2 = self.sigmoid2(self.linear2(sigmoid_out1))
        return self.softmax(self.linear3(sigmoid_out2))

net = TitanicSimpleNNModel()
criterion = nn.CrossEntropyLoss()
num_epochs = 200

optimizer = torch.optim.Adam(net.parameters(), lr=0.1)
input_tensor = torch.from_numpy(train_features).type(torch.FloatTensor)
label_tensor = torch.from_numpy(train_labels)
```

The loss function used in the model is cross entropy loss. You can choose different types of loss functions based on input data and accuracy. Since the Adam optimizer is considered to be relevant in most use cases, it is applied.

```python
for epoch in range(num_epochs):
    output = net(input_tensor)
    loss = criterion(output, label_tensor)
    optimizer.zero_grad()
    loss.backward()
    optimizer.step()
    if epoch % 20 == 0:
        print ('Epoch {}/{} => Loss: {:.2f}'.format(epoch+1, num_epochs,
        loss.item()))
torch.save(net.state_dict(), '/model.pt')

Epoch 1/200 => Loss: 0.70
Epoch 21/200 => Loss: 0.55
Epoch 41/200 => Loss: 0.50
Epoch 61/200 => Loss: 0.49
Epoch 81/200 => Loss: 0.48
Epoch 101/200 => Loss: 0.49
Epoch 121/200 => Loss: 0.47
Epoch 141/200 => Loss: 0.47
Epoch 161/200 => Loss: 0.47
```

```
Epoch 181/200 => Loss: 0.47
```

The number of iterations, 200, tries to reduce the cross entropy loss. The `torch.save` function is used to store the trained model in the default directory. You can change the path in /model path.

```
out_probs = net(input_tensor).detach().numpy()
out_classes = np.argmax(out_probs, axis=1)
print("Train Accuracy:", sum(out_classes == train_labels) / len(train_labels))
Train Accuracy: 0.8523274478330658
```

The input tensor is detached from the neural network model and converted to a numpy array in order to store the class probabilities. The training accuracy is 85%.

```
test_input_tensor = torch.from_numpy(test_features).type(torch.FloatTensor)
out_probs = net(test_input_tensor).detach().numpy()
out_classes = np.argmax(out_probs, axis=1)
print("Test Accuracy:", sum(out_classes == test_labels) / len(test_labels))
```

```
Test Accuracy: 0.832089552238806
```

The integrated gradient is extracted from the neural network model. IG can be extracted using the `attribute` function. You must make the return convergence delta True and applying it requires gradients on the test dataset.

```
ig = IntegratedGradients(net)
test_input_tensor.requires_grad_()
attr, delta = ig.attribute(test_input_tensor,target=1, return_convergence_
delta=True)
attr = attr.detach().numpy()
np.round(attr,2)
array([[-0.7 , 0.09, -0. , ..., 0. , 0. , -0.33], [-2.78, -0. , -0. , ...,
0. , 0. , -1.82], [-0.65, 0. , -0. , ..., 0. , 0. , -0.31], ..., [-0.47,
-0. , -0. , ..., 0.71, 0. , -0. ], [-0.1 , -0. , -0. , ..., 0. , 0. , -0.1
], [-0.7 , 0. , -0. , ..., 0. , 0. , -0.28]])
```

The `attr` contains the feature importance of the input features from the model.

```
importances = np.mean(attr, axis=0)
for i in range(len(feature_names)):
```

```
        print(feature_names[i], ": ", '%.3f'%(importances[i]))
Age :   -0.574
SibSp :   -0.010
Parch :   -0.026
Fare :   0.278
female :   0.101
male :   -0.460
embark_C :   0.042
embark_Q :   0.005
embark_S :   -0.021
class_1 :   0.067
class_2 :   0.090
class_3 :   -0.144
```

The importance of the input features having negative and positive numbers is the following: a negative number shows having a negative impact on the class probability and a positive one adds to the probability score. The feature importance shows how relevant a feature is in performing classification.

Recipe 10-3. Neuron Importance of a Deep Learning Model

Problem

How do you calculate the importance of neurons in a deep learning model?

Solution

The layer of conductance combines the neuron activation by taking the partial derivative of the neuron with respect to input and output. The conductance layer builds on the integrated gradients by looking at the flow of IG attribution.

How It Works

This code shows how to compute the neuron importance of a deep learning model:

```
cond = LayerConductance(net, net.sigmoid1)
```

net.sigmoid1 is the first hidden layer. net is the deep learning model object. Layer conductance is the function, and conductance values are stored in the object cond.

```
cond_vals = cond.attribute(test_input_tensor,target=1)
cond_vals = cond_vals.detach().numpy()
Average_Neuron_Importances = np.mean(cond_vals, axis=0)
Average_Neuron_Importances
array([ 0.03051018, -0.23244175, 0.04743345, 0.02102091, -0.08071412,
-0.09040915, -0.13398956, -0.04666219, 0.03577907, -0.07206058,
-0.15658873, 0.03491106], dtype=float32)
```

There are 12 neurons in the hidden layer, which is why you have 12 elements in neuron importance, and since there are many layers, it is derived as an average of conductance values.

```
neuron_cond = NeuronConductance(net, net.sigmoid1)
```

In a similar manner, the neuron conductance provides the conductance values from hidden layer.

```
neuron_cond_vals_10 = neuron_cond.attribute(test_input_tensor, neuron_
selector=10, target=1)
neuron_cond_vals_0 = neuron_cond.attribute(test_input_tensor, neuron_
selector=0, target=1)
# Average Feature Importances for Neuron 0
nn0 = neuron_cond_vals_0.mean(dim=0).detach().numpy()
np.round(nn0,3)
array([ 0.008, 0. , 0. , 0.028, 0. , -0.004, -0. , 0. , -0.001, -0. , 0. ,
-0. ], dtype=float32)
```

Recipe 10-4. Installation of Skorch
Problem

How do you install skorch?

Solution

There are two ways to install skorch: either using the sudo command or using the pip command.

How It Works

The following syntax explains how to install the library:

```
pip install -U skorch
OR

conda install -U skorch

Looking in indexes: https://pypi.org/simple, https://us-python.pkg.dev/
colab-wheels/public/simple/
Collecting skorch
  Downloading skorch-0.11.0-py3-none-any.whl (155 kB)
     |████████████████████████████████| 155 kB 27.9 MB/s
Requirement already satisfied: numpy>=1.13.3 in /usr/local/lib/python3.7/
dist-packages (from skorch) (1.21.6)
Requirement already satisfied: scikit-learn>=0.19.1 in /usr/local/lib/
python3.7/dist-packages (from skorch) (1.0.2)
Requirement already satisfied: tqdm>=4.14.0 in /usr/local/lib/python3.7/
dist-packages (from skorch) (4.64.0)
Requirement already satisfied: scipy>=1.1.0 in /usr/local/lib/python3.7/
dist-packages (from skorch) (1.7.3)
Requirement already satisfied: tabulate>=0.7.7 in /usr/local/lib/python3.7/
dist-packages (from skorch) (0.8.10)
Requirement already satisfied: joblib>=0.11 in /usr/local/lib/python3.7/
dist-packages (from scikit-learn>=0.19.1->skorch) (1.1.0)
```

```
Requirement already satisfied: threadpoolctl>=2.0.0 in /usr/local/lib/
python3.7/dist-packages (from scikit-learn>=0.19.1->skorch) (3.1.0)
Installing collected packages: skorch
Successfully installed skorch-0.11.0
```

Recipe 10-5. Skorch Components for a Neuralnet Classifier

Problem

How do you train a scorch-based neuralnet classifier?

Solution

Scorch is a Scikit-learn—compatible library that wraps PyTorch to provide functionalities for training neural networks. The advantage of this library is to reduce the boilerplate code. Skorch can be used for classification and regression. Skorch.neuralnet classifier and scorch.neuralnet regressor are major modules. The good part of the skorch module is that the model training process is fast and it displays the results in a nice way.

How It Works

The components of Scikit-learn such as fitting, preprocessing, predicting, cross validation, metrics, grid searches, and the pipeline are very popular and you will want to use and apply them on top of a neural network model that is trained using the PyTorch library because PyTorch has become a standard tool for training all sorts of deep learning models.

```
import torch
from torch import nn
import numpy as np
import torch.nn.functional as F
from sklearn.datasets import make_classification

X, y = make_classification(2000, 10, random_state=0)
X, y = X.astype(np.float32), y.astype(np.int64)
```

The above code contains standard sample data for classification.

```python
class ClassifierModule(nn.Module):
    def __init__(
            self,
            num_units=30,
            nonlin=F.relu,
            dropout=0.5,
    ):
        super(ClassifierModule, self).__init__()
        self.num_units = num_units
        self.nonlin = nonlin
        self.dropout = dropout

        self.dense0 = nn.Linear(10, num_units)
        self.nonlin = nonlin
        self.dropout = nn.Dropout(dropout)
        self.dense1 = nn.Linear(num_units, 10)
        self.output = nn.Linear(10, 2)

    def forward(self, X, **kwargs):
        X = self.nonlin(self.dense0(X))
        X = self.dropout(X)
        X = F.relu(self.dense1(X))
        X = F.softmax(self.output(X), dim=-1)
        return X

from skorch import NeuralNetClassifier
net = NeuralNetClassifier(
    ClassifierModule,
    max_epochs=20,
    lr=0.1,
#     device='cuda',  # uncomment this to train with CUDA
)
```

If you have access to the GPU, then the commented code in the classifier can be uncommented.

```
net.get_params()
net.fit(X, y)
```

epoch	train_loss	valid_acc	valid_loss	dur
1	0.7076	0.5325	0.6674	0.0255
2	0.6532	0.8050	0.5975	0.0217
3	0.5660	0.9675	0.4638	0.0206
4	0.4265	0.9800	0.2979	0.0189
5	0.2913	0.9875	0.1789	0.0207
6	0.2128	0.9925	0.1150	0.0208
7	0.1689	0.9900	0.0825	0.0207
8	0.1496	0.9900	0.0644	0.0195
9	0.1183	0.9900	0.0545	0.0197
10	0.1218	0.9900	0.0490	0.0198
11	0.1240	0.9900	0.0464	0.0203
12	0.1090	0.9900	0.0428	0.0213
13	0.1050	0.9900	0.0410	0.0215
14	0.1067	0.9875	0.0399	0.0214
15	0.1072	0.9900	0.0392	0.0211
16	0.0958	0.9900	0.0378	0.0241
17	0.0964	0.9900	0.0371	0.0213
18	0.0986	0.9925	0.0361	0.0223
19	0.0884	0.9900	0.0363	0.0205
20	0.0991	0.9900	0.0366	0.0213

```
<class 'skorch.classifier.NeuralNetClassifier'>[initialized](
  module_=ClassifierModule(
    (dense0): Linear(in_features=10, out_features=30, bias=True)
    (dropout): Dropout(p=0.5, inplace=False)
    (dense1): Linear(in_features=30, out_features=10, bias=True)
    (output): Linear(in_features=10, out_features=2, bias=True)
  ),
)
```

```
list(net.get_params())
['module', 'criterion', 'optimizer', 'lr', 'max_epochs', 'batch_
size', 'iterator_train', 'iterator_valid', 'dataset', 'train_split',
'callbacks', 'predict_nonlinearity', 'warm_start', 'verbose', 'device',
'_kwargs', 'classes', 'callbacks__epoch_timer', 'callbacks__train_loss',
'callbacks__train_loss__name', 'callbacks__train_loss__lower_is_better',
'callbacks__train_loss__on_train', 'callbacks__valid_loss', 'callbacks__
valid_loss__name', 'callbacks__valid_loss__lower_is_better', 'callbacks__
valid_loss__on_train', 'callbacks__valid_acc', 'callbacks__valid_acc__
scoring', 'callbacks__valid_acc__l
lower_is_better', 'callbacks__valid_acc__on_train', 'callbacks__valid_
acc__name', 'callbacks__valid_acc__target_extractor', 'callbacks__valid_
acc__use_caching', 'callbacks__print_log', 'callbacks__print_log__keys_
ignored', 'callbacks__print_log__sink', 'callbacks__print_log__tablefmt',
'callbacks__print_log__floatfmt', 'callbacks__print_log__stralign']
```

Once the model is trained, you can use the predict function from Scikit-learn to generate the predictions and probability function to get the class probabilities.

```
y_pred = net.predict(X[:5])
y_pred
array([1, 0, 0, 1, 1])

y_proba = net.predict_proba(X[:5])
y_proba
array([[7.7738642e-04, 9.9922264e-01], [9.9628782e-01, 3.7122301e-03],
[9.9648917e-01, 3.5108225e-03], [3.2411060e-01, 6.7588937e-01],
[4.5940662e-03, 9.9540591e-01]], dtype=float32)
```

Recipe 10-6. Skorch Neuralnet Regressor
Problem

How do you train a regression model using skorch?

Solution

Regression model training follows the standard practice of using a neural network model. Here you are going to use one to make the regressor function generate some synthetic data and use the skorch functions to train the model.

How It Works

The following syntax explains how to execute this:

```
from sklearn.datasets import make_regression
X_regr, y_regr = make_regression(1000, 20, n_informative=10,
random_state=0)
X_regr = X_regr.astype(np.float32)
y_regr = y_regr.astype(np.float32) / 100
y_regr = y_regr.reshape(-1, 1)
X_regr.shape, y_regr.shape, y_regr.min(), y_regr.max()
((1000, 20), (1000, 1), -6.4901485, 6.154505)

class RegressorModule(nn.Module):
    def __init__(
            self,
            num_units=10,
            nonlin=F.relu,
    ):
        super(RegressorModule, self).__init__()
        self.num_units = num_units
        self.nonlin = nonlin

        self.dense0 = nn.Linear(20, num_units)
        self.nonlin = nonlin
```

```python
        self.dense1 = nn.Linear(num_units, 10)
        self.output = nn.Linear(10, 1)

    def forward(self, X, **kwargs):
        X = self.nonlin(self.dense0(X))
        X = F.relu(self.dense1(X))
        X = self.output(X)
        return X

from skorch import NeuralNetRegressor
net_regr = NeuralNetRegressor(
    RegressorModule,
    max_epochs=20,
    lr=0.1,
#     device='cuda',  # uncomment this to train with CUDA
)
net_regr.fit(X_regr, y_regr)
```

epoch	train_loss	valid_loss	dur
1	4.3247	3.0078	0.0170
2	1.7262	0.6808	0.0123
3	0.6510	0.2147	0.0115
4	0.1811	0.2132	0.0118
5	0.1906	0.1127	0.0108
6	0.1143	0.3361	0.0204
7	0.3835	0.0899	0.0113
8	0.0845	0.1574	0.0117
9	0.1099	0.0486	0.0130
10	0.0485	0.0974	0.0128
11	0.0907	0.0447	0.0108
12	0.0481	0.0947	0.0129
13	0.0881	0.0322	0.0128
14	0.0323	0.0599	0.0117
15	0.0461	0.0180	0.0115
16	0.0161	0.0328	0.0123

17	0.0231	0.0125	0.0123
18	0.0098	0.0208	0.0123
19	0.0143	0.0102	0.0112
20	0.0074	0.0153	0.0121

```
<class 'skorch.regressor.NeuralNetRegressor'>[initialized](
  module_=RegressorModule(
    (dense0): Linear(in_features=20, out_features=10, bias=True)
    (dense1): Linear(in_features=10, out_features=10, bias=True)
    (output): Linear(in_features=10, out_features=1, bias=True)
  ),
)

y_pred = net_regr.predict(X_regr[:5])
y_pred
array([[ 0.7368696 ], [-1.2884711 ], [-0.51758516], [-0.11890286],
[-0.61254007]], dtype=float32)
```

Recipe 10-7. Skorch Model Save and Load

Problem

How do you save and load a model object generated by skorch?

Solution

Using the pickle library for storing the model objects as serialized objects and loading these objects into another environment is done by using the Scikit-learn library. Here you are going to use the skorch library to save and load the model.

How It Works

The following syntax explains how to execute this:

```
import pickle
file_name = '/tmp/mymodel.pkl'
with open(file_name, 'wb') as f:
    pickle.dump(net, f)
```

```
with open(file_name, 'rb') as f:
    new_net = pickle.load(f)

net.save_params(f_params=file_name)  # a file handler also works
```

If you store the model object as saved parameters of the model, then you need to initialize the model again and assign that to a new object.

```
# first initialize the model
new_net = NeuralNetClassifier(
    ClassifierModule,
    max_epochs=20,
    lr=0.1,
).initialize()

new_net.load_params(file_name)
```

Recipe 10-8. Skorch Model Pipeline Creation

Problem

How do you create a pipeline for neural network models using skorch?

Solution

A pipeline object is a structure where a series of operations can be scheduled in a process so that the model training and execution happens in a sequential manner.

How It Works

The following script shows how this can be done using scorch:

```
from sklearn.pipeline import Pipeline
from sklearn.preprocessing import StandardScaler
pipe = Pipeline([
    ('scale', StandardScaler()),
```

```
    ('net', net),
])
pipe.fit(X, y)
Re-initializing module.
Re-initializing criterion.
Re-initializing optimizer.
```

epoch	train_loss	valid_acc	valid_loss	dur
1	0.6925	0.5250	0.6640	0.0355
2	0.6361	0.9075	0.5834	0.0354
3	0.5447	0.9550	0.4427	0.0340
4	0.4197	0.9675	0.2898	0.0207
5	0.3019	0.9775	0.1798	0.0214
6	0.2282	0.9825	0.1206	0.0217
7	0.1790	0.9875	0.0869	0.0207
8	0.1550	0.9875	0.0697	0.0208
9	0.1473	0.9875	0.0594	0.0196
10	0.1249	0.9875	0.0525	0.0200
11	0.1294	0.9900	0.0482	0.0213
12	0.1194	0.9925	0.0446	0.0220
13	0.1192	0.9950	0.0428	0.0291
14	0.1035	0.9925	0.0406	0.0215
15	0.0989	0.9925	0.0394	0.0223
16	0.0999	0.9925	0.0386	0.0206
17	0.0928	0.9925	0.0376	0.0200
18	0.0980	0.9925	0.0370	0.0189
19	0.0969	0.9925	0.0364	0.0211
20	0.0876	0.9925	0.0358	0.0213

```
Pipeline(steps=[('scale', StandardScaler()), ('net', <class 'skorch.
classifier.NeuralNetClassifier'>[initialized]( module_=ClassifierModule
( (dense0): Linear(in_features=10, out_features=30, bias=True) (dropout):
Dropout(p=0.5, inplace=False) (dense1): Linear(in_features=30, out_
features=10, bias=True) (output): Linear(in_features=10, out_features=2,
bias=True) ), ))])
```

```
y_proba = pipe.predict_proba(X[:5])
y_proba
array([[0.00224374, 0.9977563 ], [0.9986193 , 0.00138069], [0.99899906,
0.00100095], [0.30393705, 0.6960629 ], [0.00816792, 0.9918321 ]],
dtype=float32)
```

Recipe 10-9. Skorch Model Epoch Scoring

Problem

How do you use callbacks in neural network models using skorch?

Solution

While training deep learning models, you can leverage the callback function to do epoch scoring. This requires a scoring function that need to be defined. After completion of each epoch, the function needs to be called in and, if the desired level of accuracy is achieved, then it should be highlighted.

How It Works

The following code shows how:

```
from skorch.callbacks import EpochScoring
auc = EpochScoring(scoring='roc_auc', lower_is_better=False)
net = NeuralNetClassifier(
    ClassifierModule,
    max_epochs=20,
    lr=0.1,
    callbacks=[auc],
)

net.fit(X, y)
```

epoch	roc_auc	train_loss	valid_acc	valid_loss	dur
1	0.9544	0.6614	0.8900	0.6294	0.0193
2	0.9845	0.5875	0.9625	0.5233	0.0208
3	0.9899	0.4798	0.9800	0.3647	0.0198
4	0.9945	0.3600	0.9825	0.2302	0.0208
5	0.9972	0.2682	0.9850	0.1451	0.0200
6	0.9975	0.2087	0.9850	0.1002	0.0187
7	0.9978	0.1869	0.9850	0.0762	0.0191
8	0.9979	0.1699	0.9850	0.0640	0.0222
9	0.9980	0.1430	0.9875	0.0567	0.0200
10	0.9981	0.1338	0.9875	0.0500	0.0201
11	0.9981	0.1214	0.9875	0.0464	0.0355
12	0.9981	0.1167	0.9900	0.0442	0.0346
13	0.9982	0.1072	0.9875	0.0419	0.0396
14	0.9981	0.1152	0.9900	0.0404	0.0337
15	0.9981	0.1086	0.9900	0.0395	0.0341
16	0.9982	0.0905	0.9875	0.0387	0.0338
17	0.9981	0.0983	0.9875	0.0382	0.0466
18	0.9981	0.0929	0.9875	0.0373	0.0618
19	0.9982	0.1009	0.9875	0.0368	0.0302
20	0.9982	0.0981	0.9875	0.0362	0.0219

```
<class 'skorch.classifier.NeuralNetClassifier'>[initialized]
( module_=ClassifierModule( (dense0): Linear(in_features=10, out_
features=30, bias=True) (dropout): Dropout(p=0.5, inplace=False) (dense1):
Linear(in_features=30, out_features=10, bias=True) (output): Linear
(in_features=10, out_features=2, bias=True) ), )
```

```
print(', '.join(net.prefixes_))
iterator_train, iterator_valid, callbacks, dataset, module, criterion,
optimizer
```

Recipe 10-10. Grid Search for Best Hyper Parameter

Problem

How do you use a grid search for hyper parameter training using skorch?

Solution

The hyper parameter values in a deep learning model may produce multiple models. You need to apply a logic to find out which combinations of hyper parameters produce the best model; hence they can be called the best hyper parameters. The optimal hyper parameter is likely to vary based on the maximum epochs specified.

How It Works

The following script shows how to do so:

```
from sklearn.model_selection import GridSearchCV
net = NeuralNetClassifier(
    ClassifierModule,
    max_epochs=20,
    lr=0.1,
    optimizer__momentum=0.9,
    verbose=0,
    train_split=False,
)

params = {
    'lr': [0.05, 0.1],
    'module__num_units': [10, 20],
    'module__dropout': [0, 0.5],
    'optimizer__nesterov': [False, True],
}

gs = GridSearchCV(net, params, refit=False, cv=3, scoring='accuracy',
verbose=2)
```

```
gs.fit(X, y)
Fitting 3 folds for each of 16 candidates, totalling 48 fits
[CV] END lr=0.05, module__dropout=0, module__num_units=10,
optimizer__nesterov=False; total time=    0.4s
[CV] END lr=0.05, module__dropout=0, module__num_units=10,
optimizer__nesterov=False; total time=    0.3s
[CV] END lr=0.05, module__dropout=0, module__num_units=10,
optimizer__nesterov=False; total time=    0.3s
[CV] END lr=0.05, module__dropout=0, module__num_units=10,
optimizer__nesterov=True; total time=    0.3s
[CV] END lr=0.05, module__dropout=0, module__num_units=10,
optimizer__nesterov=True; total time=    0.3s
[CV] END lr=0.05, module__dropout=0, module__num_units=10,
optimizer__nesterov=True; total time=    0.3s
[CV] END lr=0.05, module__dropout=0, module__num_units=20,
optimizer__nesterov=False; total time=    0.3s
[CV] END lr=0.05, module__dropout=0, module__num_units=20,
optimizer__nesterov=False; total time=    0.3s
[CV] END lr=0.05, module__dropout=0, module__num_units=20,
optimizer__nesterov=False; total time=    0.3s
[CV] END lr=0.05, module__dropout=0, module__num_units=20,
optimizer__nesterov=True; total time=    0.3s
[CV] END lr=0.05, module__dropout=0, module__num_units=20,
optimizer__nesterov=True; total time=    0.3s
[CV] END lr=0.05, module__dropout=0, module__num_units=20,
optimizer__nesterov=True; total time=    0.3s
[CV] END lr=0.05, module__dropout=0.5, module__num_units=10,
optimizer__nesterov=False; total time=    0.3s
[CV] END lr=0.05, module__dropout=0.5, module__num_units=10,
optim...................

print(gs.best_score_, gs.best_params_)
.988499744121933 {'lr': 0.1, 'module__dropout': 0.5, 'module__num_units':
20, 'optimizer__nesterov': False}
```

Conclusion

This chapter provided options to include Scikit-learn—compatible functions and build a wrapper that can be added on top of PyTorch-based models and can act as a standard Scikit-learn—based model. This chapter also included recipes on model interpretability, which is very important for any deep learning model used for supervised learning-related tasks such as regression and classification.

Index

A, B

Activation functions
    bilinear transformation, 119
    forward/backward
        propagation, 142–145
    hyperbolic tangent function, 121
    leaky ReLU, 124
    linear function, 118
    log sigmoid transfer function, 122
    mathematical function, 118
    matplotlib, 125
    rectified linear unit (ReLU), 123
    sequential module, 159
    sigmoid function, 120, 127
    softplus, 128
    tanh, 127
    TensorFlow, 118
    visualization, 125
    working process, 117, 125–128
Application programming
        interface (API), 159
Artificial neural network (ANN), 117
Audio/image processing tasks, 213
    filter bank, 224–226
    Griffin-Lim algorithm (GLA), 223–225
    librosa mel scale
        conversion, 226–229
    librosa installation, 219–221
    loading data, 218, 219
    MFCC/LFCC modules, 229–232
    spectogram, 213–215, 221, 222
    torchaudio module, 216, 217, 227

C

Captum
    installation, 238, 239
    primary attribution, 239–244
Central processing units (CPUs), 1, 28
Computational linguistics, *see* Natural
        language processing
Continuous bag of words (CBOW), 178–181
Convolutional neural network (CNN),
        29, 49, 157
    convolution layer, 71–74
    dataset, 70, 71
    hyperparameters, 69
    implementation, 69–77
    MNIST dataset, 70
    reloading model, 77–80
    visualization, 74

D, E

Deep learning models, 157
    distributed torch architectures, 188
    neuron activation, 244, 245
    overfitting, 105–108

Autoencoders function
    architecture, 94
    fine-tune results, 98–101
    hyperbolic tangent function, 97, 98
    testing, 95
    torchvision library, 94–97
    working process, 93

P. Mishra, *PyTorch Recipes*, https://doi.org/10.1007/978-1-4842-8925-9

Printed in the United States
by Baker & Taylor Publisher Services